Bodybuilding Anatomy

Bodybuilding Anatomy

Michael Israetel, PhD
Jared Feather, MS
Christle Guevarra, DO

Library of Congress Cataloging-in-Publication Data

Names: Israetel, Michael, 1984- author. | Feather, Jared, 1993- author. | Guevarra, Christle, 1982- author.
Title: Bodybuilding anatomy / Michael Israetel, Jared Feather, Christle Guevarra.
Description: Champaign, IL : Human Kinetics, 2026.
Identifiers: LCCN 2024037101 (print) | LCCN 2024037102 (ebook) | ISBN 9781718225985 (paperback) | ISBN 9781718225992 (epub) | ISBN 9781718226005 (pdf)
Subjects: LCSH: Bodybuilding--Training. | Bodybuilding--Physiological aspects.
Classification: LCC GV546.5 .I78 2026 (print) | LCC GV546.5 (ebook) | DDC 613.7/13--dc23/eng/20241118
LC record available at https://lccn.loc.gov/2024037101
LC ebook record available at https://lccn.loc.gov/2024037102

ISBN: 978-1-7182-2598-5 (print)

Senior Acquisitions Editors: Korey Van Wyk and Michelle Earle; **Senior Developmental Editor:** Cynthia McEntire; **Managing Editor:** Kevin Matz; **Copyeditor:** Jackie Gibson; **Senior Graphic Designer:** Julie L. Denzer; **Cover Designer:** Keri Evans; **Cover Design Specialist:** Susan Rothermel Allen; **Photographs (used as anatomical reference):** Tay Price Photography Corp.; **Photo Production Specialist:** Amy M. Rose; **Senior Art Manager:** Kelly Hendren; **Illustrations:** © Human Kinetics; **Printer:** Versa Press

We thank Dragon's Lair Gym in Las Vegas, Nevada, for assistance in providing the location for the photo shoot for this book.

Human Kinetics books are available at special discounts for bulk purchase. Special editions or book excerpts can also be created to specification. For details, contact the Special Sales Manager at Human Kinetics.

Printed in the United States of America 10 9 8 7 6 5 4 3 2 1

The paper in this book is certified under a sustainable forestry program.

Human Kinetics
1607 N. Market Street
Champaign, IL 61820
USA

United States and International
Website: **US.HumanKinetics.com**
Email: info@hkusa.com
Phone: 1-800-747-4457

Canada
Website: **Canada.HumanKinetics.com**
Email: info@hkcanada.com

E9305

CONTENTS

INTRODUCTION

Bodybuilding Anatomy is for anyone interested in optimizing their training, with the ultimate goal of transforming their physique into one that is both leaner and more muscular. No one wants to add muscle randomly, with exercises and techniques that could be inefficient or injurious. Ideally, we would like to add muscle to the specific parts of our bodies where we think it will look best, using methods that are as safe and effective as possible.

Our aim in this book is to highlight the dynamic relationship between muscle structure and function and bodybuilding success. Each chapter will build your knowledge from the ground up. We aim to give you a comprehensive understanding of bodybuilding anatomy that is applicable in the weight room so you can achieve the best results. To help guide you, each exercise will be accompanied by an anatomical illustration that demonstrates how each exercise is performed. These illustrations are color coded to indicate the primary and secondary muscles and the connective tissues featured in each exercise.

Chapter 1 starts with the chest, which plays a huge role in upper body aesthetics. We've included exercises to provide maximal tension to the pectorales, such as the cambered bar bench, deficit push-up, dumbbell press flye, and incline cable flye. Between the sizable exercise list and the safety sections, you will not only be able to build a chest that stands out but also decrease your risk of injury.

Chapter 2 continues to build upper body aesthetics, turning the focus to the shoulders. We've included a few favorite exercises that work various parts of the deltoids to build some impressive capped delts. These include the super-ROM dumbbell lateral raise, freemotion Y-lateral raise, one-arm rear cable delt flye, and cable face-pull. By keeping the emphasis on proper technique and safety, you can build your shoulders without sacrificing your joint health.

Chapter 3 swings around to the back and walks you through how to build a back that's thick and wide, giving that desired V-shaped appearance onstage. We have included several pulling exercises along with variations for each. This will give you more than enough ways to build impressive trapezius muscles, latissimus dorsi muscles, and spinal erectors.

Chapter 4 is the last of the chapters covering the upper body, with emphasis on the arms. We will help you build a better understanding of the muscles of the forearms as well as your triceps and biceps. We've also offered a large variety of exercises to choose from.

Chapter 5 moves away from the upper body to the torso. Exercises such as the V-up or cable crunch will target the abdominals effectively, creating that highly coveted "six pack."

Chapter 6 focuses on the legs. This chapter is packed with an array of exercises that work the gluteal muscles, quadriceps muscles, hamstrings, and calves. This will allow you to build a huge lower body to match the upper body.

Chapter 7 explains how you can choose exercises in a systematic fashion to match your needs and goals. You will learn how to use the stimulus-to-fatigue ratio (SFR) to choose exercises as well as how to maximize SFR to make the most gains—without sacrificing joint health and generating needless fatigue.

Lastly, chapter 8 shows some of the posing basics seen in bodybuilding competition, to help you start showing off those hard-earned muscles, whether it is onstage or in front of your mirror at home.

Let's dive in and bring you up to date on the latest ways to target the muscles you want to grow.

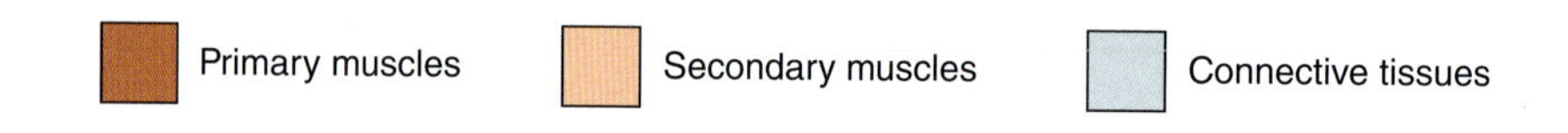

CHAPTER 1

Chest

The chest is a key focal point in bodybuilding, because it significantly affects overall upper body aesthetics and overall strength. A well-developed chest not only creates a powerful and visually striking physique, but it also plays a role in various athletic movements (such as pushing). This chapter will provide a thorough overview of exercises that effectively target this region.

The anterior chest wall is composed of several important muscles, including the pectoralis major, pectoralis minor, and serratus anterior (figure 1.1). Collectively, these muscles contribute to the overall stability of the scapula (shoulder blade) and the glenohumeral joint (shoulder joint).

When working on chest development, the main muscle to target is the pectoralis major: a large, fan-shaped muscle that comprises two main sections, or heads. The clavicular head, which makes up the upper part of the chest, originates from the clavicle (collarbone). The sternal head, which makes up the lower part of the chest, mostly originates from the sternum (breastbone) and also includes the cartilage of the first six to seven ribs as well as the aponeurosis (sheet of connective tissues) of the external oblique muscle. Both muscle heads move outward across the chest and converge at a single tendon that inserts into the humerus (upper arm bone). The pectoralis major is mainly responsible for shoulder flexion and horizontal adduction, as well as internal rotation of the arm. The best way to target the pectoralis major is with exercises that involve moving the arm away from the body or across the chest (such as a bench press or dumbbell flye, respectively).

The serratus anterior muscle is also part of the chest, albeit a much smaller part. The serratus anterior originates from the upper eight or nine ribs and inserts into the medial border of the scapula. Its main function is to protract the scapula forward and away from the

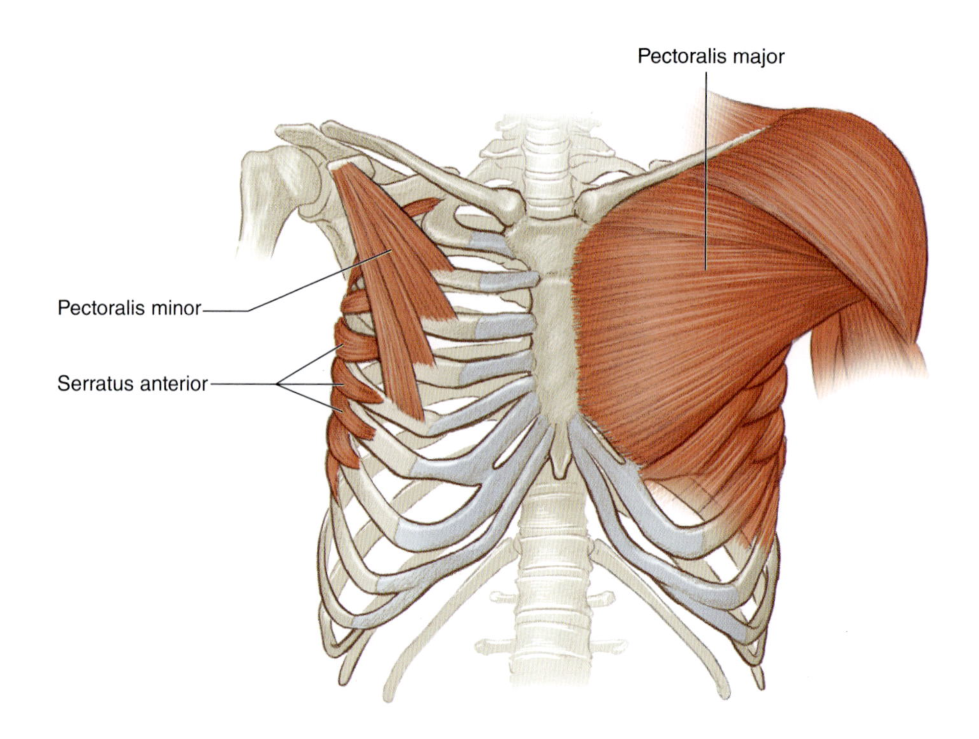

Figure 1.1 Chest.

spine, which is critical for pushing exercises. Given its minuscule role in the appearance of the chest, most of the time it is unnecessary to specifically target this muscle for the purposes of bodybuilding.

The pectoralis minor is a very small, triangular-shaped muscle that sits underneath the pectoralis major. Its main function is to provide stability to the scapula. Similar to the serratus anterior, given its minuscule role in the appearance of the chest, most bodybuilding programs do not target the pectoralis minor.

FLAT BARBELL BENCH PRESS

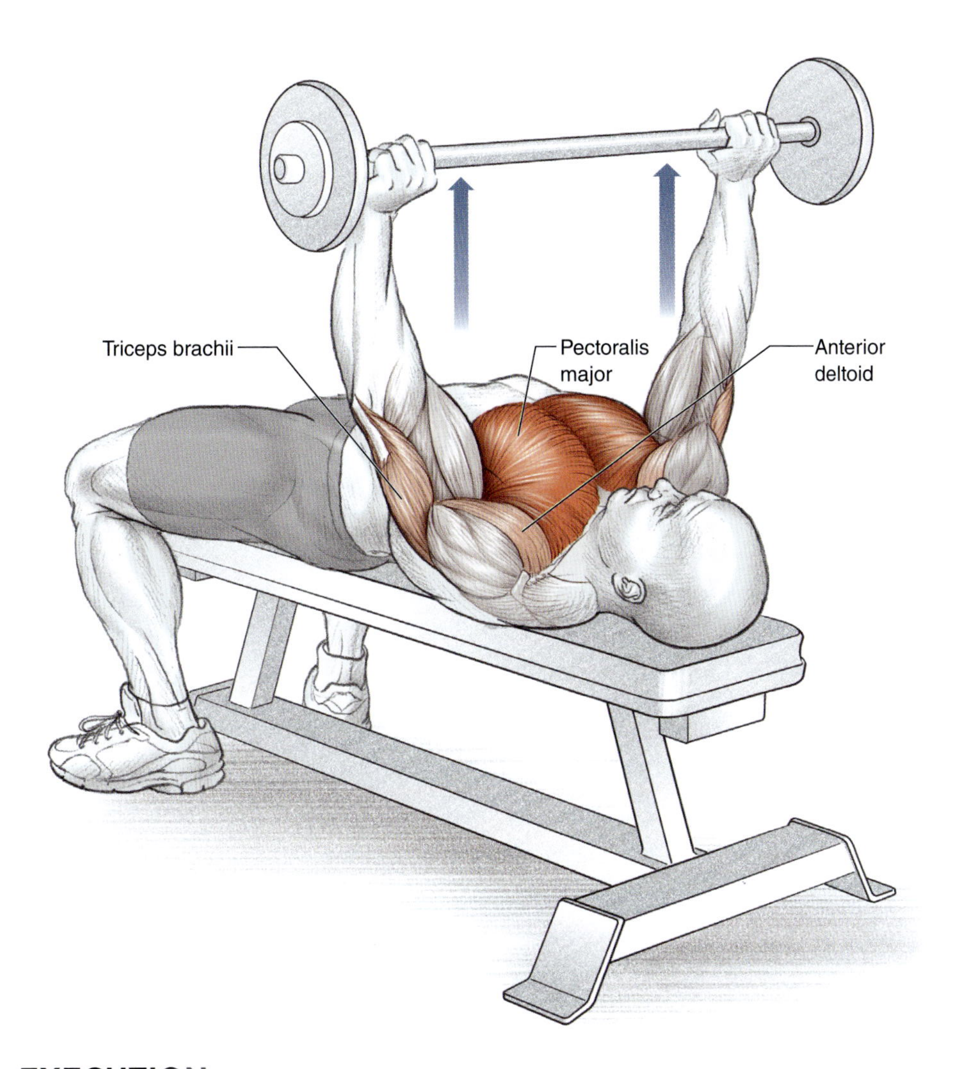

EXECUTION

1. Start by lying supine on the flat bench. Grip the barbell with your thumbs wrapped around the bar, and unrack the barbell. Begin with the barbell fully extended at lockout, with your scapulae retracted or depressed (not shrugged up toward the ears) and your spine in a slight lordotic (arched) position. Your glutes should never leave the bench, and your feet should be comfortably and firmly planted on the floor for stability. If your feet do not reach the floor, you may keep your toes planted on the floor, or you may consider putting a weight plate underneath your feet.
2. Slowly and with control, loosen your elbows and bring the bar down toward your body until it gently touches the lower part of your chest (anywhere between the bottom of the sternum and just above the nipple line, depending on comfort).

> *continued*

3. Hold for about one second, then press the load back up to lockout while keeping your scapulae back the whole time.
4. Throughout execution, keep the slight lordotic posture, and focus on using the pushing musculature to move the load as the body remains rigid.

MUSCLES INVOLVED

Primary: Pectoralis major

Secondary: Anterior deltoid, triceps brachii

EXERCISE NOTES

As you descend, try to reach your chest up to meet the bar on its way down. This will help you stretch and stimulate the pectorales (pecs) better, so long as your glutes never leave the bench.

You may have heard people talk about which angle of humeral abduction and adduction is ideal. In reality, there is no wrong angle as long as you feel strong when holding the position and you aren't creating pain in your elbow and shoulder joints. It will work just fine to keep your elbows flush to your sides, keep them abducted way out, or hold them anywhere in between. In fact, these differences could work well as variations. We recommend first keeping the upper arm at about a 45-degree angle to the torso. Once you are comfortable with the movement, you can experiment with smaller and larger angles of adduction.

TIMING CONSIDERATIONS

Because this exercise requires a lot of stabilization and control, we recommend doing fewer than 20 reps. When you must maintain an arched lower back and retracted scapulae for longer amounts of time, it can become difficult to just hold that position and you may lose your ability to focus on the pressing movement that stimulates the pecs. In most cases it is best to use a range of 5 to 15 repetitions.

SAFETY

Avoid letting the weight drop to your chest, rapidly reversing at the bottom, or failing to control your scapulae positions, because these actions may result in a higher chance of injury. For maximum safety, we highly recommend a spotter. The spotter can help you unrack the weight. They should only grab the weight if you can no longer lock out a repetition and risk getting stuck. The spotter is not there to help you lift the weight during a normal repetition—that's your job!

VARIATIONS

If you grip the bar with your hands closer together and keep your elbows flush to your sides, this exercise will engage the triceps, though it will still heavily involve the pecs. If you take a wider grip and let your elbows flare up and out, the exercise will engage the triceps less but may feel strange on your shoulder joints. There are no wrong ways to perform these variations, just better and worse candidates for your programming depending on what you want to get out of the exercise. If your time is constrained and you don't want to have to work the triceps separately, the closer grip is a great option. If you'd like to focus on the triceps separately and you can achieve a great pec stimulus with minimal joint discomfort, then the wide grip version is a great choice.

INCLINE BARBELL PRESS

EXECUTION

1. Start by lying supine on the incline bench. Plant your feet firmly on the ground for maximum stability. If your feet do not reach the floor, you may keep your toes planted on the floor, or you may consider putting a weight plate underneath your feet. Arch your back a bit and retract your scapulae. Grasp the barbell, with your hands placed slightly wider than shoulder-width apart, and unrack the barbell.
2. Slowly and with control, bring the bar down to gently touch your upper chest, pausing for one second.

3. For the concentric phase, press up and slightly back so that the bar returns to the starting position in line with your chin or eyes, not just above your upper chest. The bar should make somewhat of an arcing motion as it descends and ascends.

MUSCLES INVOLVED

Primary: Upper (clavicular) pectoralis major

Secondary: Anterior deltoid, triceps brachii, lower (sternal) pectoralis major

EXERCISE NOTES

It is tempting to let your chest cave in and let the scapulae protract as you lock out each repetition, but this will degrade your initial setup as the set goes on. Instead, focus on keeping your scapulae even at lockout. If you find yourself out of position, take a moment at lockout to retract your scapulae, arch your back, and push your chest up before continuing. This will help ensure you get the most out of the movement each time. Ideally, you should avoid moving out of position in an exercise—but if you do, you should fix it and continue.

Where the bar touches your chest will depend on force production, comfort, and individual limb proportions. You want to touch where you feel strongest coming out of the hole (touching too far down the chest or too close to the clavicle can reduce strength substantially); where your shoulder joints feel the best and your pecs feel a deep stretch (for contraction); and wherever your individual limb ratios seem to place the bar by default if you just come down slowly without aiming for a certain point.

TIMING CONSIDERATIONS

Because the incline barbell press targets the upper pecs, you can use it to grow your upper pecs specifically. It will also stimulate the middle of the pecs and overall growth of the pecs. This exercise may not be the best choice if you want to specifically target your lower pecs.

SAFETY

If you often protract your scapulae during this exercise by letting your initial position break down, you may risk a slightly higher chance of shoulder joint discomfort. Try to maintain and reassert a slightly arched back and retracted scapulae.

VARIATIONS

As with the flat barbell bench press, you can adjust to a close or wide grip, depending on how much you want to engage the triceps.

LOW INCLINE CAMBERED BAR PRESS

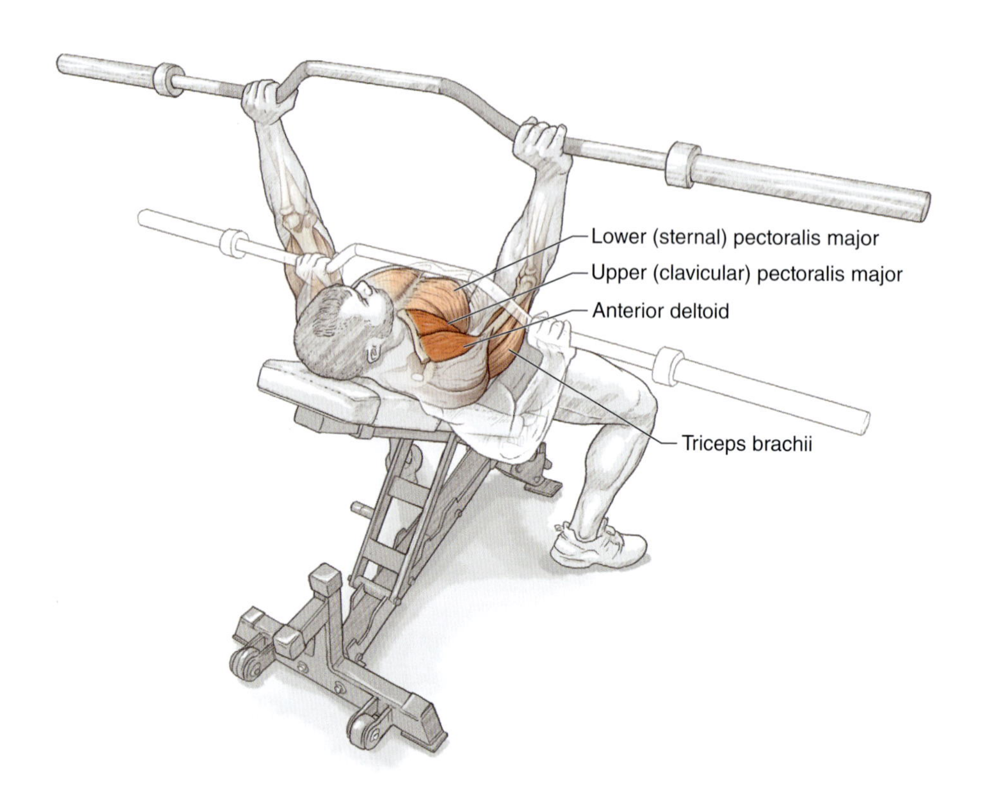

EXECUTION

1. Set the incline on the adjustable incline bench to the lowest notch (with most benches this will be around 15 or 30 degrees).
2. Start by lying supine on the incline bench. If your feet do not reach the floor, you may keep your toes planted on the floor, or you may consider putting a weight plate underneath your feet.
3. Grab a cambered bar on either side of the camber (not in the middle of the bar where the camber is). As you grab the bar and unrack it, position the camber so that it's facing away from your body and is directly vertical.
4. With your back slightly arched, and your scapulae pinched back as far as they'll go, slowly allow the bar to descend to somewhere between your lower sternum and your nipple line (depending on where you feel most comfortable in your shoulder joints and where you get the most flexibility for the deepest stretch).
5. Continue to lower the bar until you cannot stretch any lower while maintaining good technique (do not round the back or protract your scapulae). It does not matter whether the bar touches the chest. Pause for a second, then accelerate the bar back up.

MUSCLES INVOLVED

Primary: Upper (clavicular) pectoralis major, anterior deltoid
Secondary: Triceps brachii, lower (sternal) pectoralis major

EXERCISE NOTES

This exercise is all about the deep, loaded stretch, so it is important to put your hands in a position that is most comfortable for your shoulder joints and allows for the deepest stretch of the pecs. For most people this will mean a relatively close grip, but you may have to play around with it to find your best position.

As you reach close to failure on any given set of this exercise, you'll notice two distinct areas of difficulty or sticking points. The first often occurs just above the chest, and the second occurs just shy of lockout. If you're running into such a sensation, don't worry—you're not doing anything wrong. The cambered bar press is just a really gnarly exercise!

TIMING CONSIDERATIONS

This exercise is a press, so it's highly loadable and, as a barbell press, stable. Additionally, the camber makes it an incredibly stretch-biased exercise. Given these factors, this exercise can be one of the most stimulative chest exercises. Research from electromyography (EMG) results shows the low bench angle used in this exercise maximally stimulates the pecs in a pressing motion.* Because this exercise could be so stimulative and fatiguing, we recommend saving it for the end of muscle gain or fat loss blocks, where you need the most effect possible. Avoid starting this exercise directly after an active rest or maintenance phase, because just one set might leave you too sore to be able to train your pecs well again later that week.

SAFETY

When lowering the bar, you don't have to touch it to your chest if you can no longer descend further without rounding your back or letting your scapulae protract. Depending on your individual flexibility and limb length ratios, the bar may stop a few inches before the camber touches the chest. In that case, do your best to keep the technique from falling apart, and go as low as you can. Over time, even set to set, you might find yourself able to go lower while maintaining good technique—that's not mandatory, but it's great if it happens.

Pausing the bar at the bottom position for a second is a good idea in most cases, because a rapid return might stress the connective tissues of the pecs. The key is to accelerate the bar gently, moving slowly and safely from the bottom position and then moving quicker as the bar ascends and the pecs are no longer stretched so deeply.

> *continued*

VARIATIONS

If you don't have a cambered bar, you can somewhat replicate the deep stretch of this exercise by doing deficit push-ups (page 14) or by switching to dumbbells. The dumbbell press variation here is the exact same setup as the incline dumbbell press—except you should set the adjustable incline bench to 15 to 30 degrees. As with the incline dumbbell press, adjusting the incline bench to 15 to 30 degrees can help maximally stimulate the pecs.

*Rodríguez-Ridao, David, José A Antequera-Vique, Isabel Martín-Fuentes, and José M Muyor. "Effect of Five Bench Inclinations on the Electromyographic Activity of the Pectoralis Major, Anterior Deltoid, and Triceps Brachii during the Bench Press Exercise." *International Journal of Environmental Research and Public Health* 17, no. 19 (2020): 7339. doi:10.3390/ijerph17197339.

INCLINE DUMBBELL PRESS

EXECUTION

1. Set the incline bench to about 45 degrees.
2. Start by lying supine on the incline bench. Begin with the dumbbells straight overhead in lockout, your feet firmly planted on the floor, and your back slightly arched for stability. If your feet do not reach the floor, you may keep your toes planted on the floor, or you may consider putting a weight plate underneath your feet.
3. As you unlock your elbows and lower the dumbbells, it is important to bring the dumbbells out as well. When the dumbbells are at the bottom position, the inner part of the dumbbell should be as close as possible to touching only the front deltoid, and not close enough to the midline of your body to touch the pecs.

> *continued*

Incline Dumbbell Press > *continued*

4. As you begin the eccentric phase, raise your chest (while keeping your glutes on the bench), to enhance the stretch.
5. Pause for a second once the dumbbells are resting on your front deltoids (delts), then return the dumbbells back up by performing the press or concentric phase. Begin the next rep in the same fashion.

MUSCLES INVOLVED

Primary: Upper (clavicular) pectoralis major, anterior deltoid

Secondary: Triceps brachii, lower (sternal) pectoralis major

EXERCISE NOTES

It will be tempting to bring the dumbbells down to touch the chest. While this can improve the lever arm at the shoulder joint and allow you to lift more weight, it can also rob the pecs of their best potential stimulus by preventing them from being maximally stretched under load. By reaching out as well as down, you can touch the front delts or even the proximal portion of your biceps at the bottom of the press, stretching and stimulating the pecs to grow substantially more. You won't be able to lift as much weight this way, but you also won't experience as much fatigue or wear and tear on your joints.

As the dumbbells descend, try to reach the middle of your chest up to the ceiling. This will cause more stretch-mediated hypertrophy (stretching the muscles at longer length) and potentially help you establish a deeper mind–muscle connection with the pecs.

How you control and feel the bottom portion of this exercise is much more important to growth than how the contraction feels at lockout, so assign your attention mostly (if not wholly) to the bottom portion.

TIMING CONSIDERATIONS

The incline dumbbell press is the workhorse of the chest exercise world. It's almost never the wrong answer. If you do this exercise after triceps in the same session, however, you may find that the triceps are the limiting factor and become the most stimulated muscle in the exercise (leaving your pecs with a very minor stimulus). If you choose to do triceps work and a dumbbell press in the same session, remember that order matters.

SAFETY

If something goes wrong on dumbbell presses and you don't have a spotter, you can nearly always bail by letting the dumbbells fall out and away from you. That being said, rushing the eccentric phase or getting wiggly and out of position on the concentric phase will create needless risk. Because dumbbells have so many degrees of freedom (unlike barbells and especially unlike machines), it is important to focus on the stability of this movement, smoothly executing the descent and ascent. It is not worth becoming unstable just to eke out an extra rep or two. Technique comes first; rep count comes second.

VARIATIONS

To target the front delts more, reduce the adduction angle of the humerus by bringing your arms closer to your sides at the bottom. You can increase front delt and upper pec involvement by raising the bench angle to 60 degrees or more. Conversely, you can reduce front delt involvement and maximize stimulus of the pecs by adducting your elbows farther out from your body's midline and by lowering the bench angle to 15 or 30 degrees. Choose a technique that you are comfortable with, that engages the muscles you want to target, and that's easiest on the joints. You can always tweak it later to see if another variation works better for you.

DEFICIT PUSH-UP

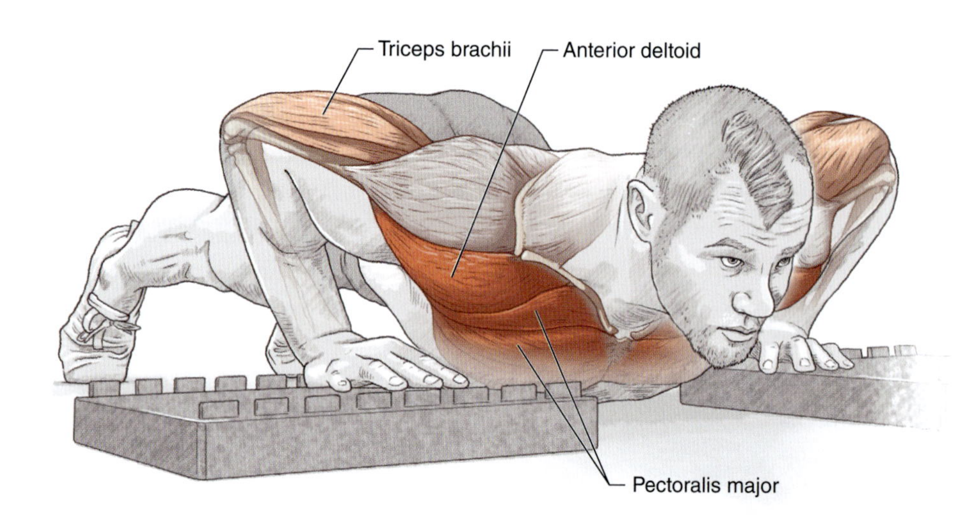

EXECUTION

1. Place a set of push-up bars or blocks onto the ground. Space them at least shoulder-width apart.
2. Get into push-up position, with your feet comfortably back so that when you descend, your shoulders pass directly over your palms. Your hands should be placed on the blocks, your feet should be close together, and your entire body should be straight.
3. Keep your upper and lower body totally rigid as you descend, with control, until your pecs are maximally stretched at the bottom.
4. Once at the bottom, gently ease yourself out of the deep stretch and accelerate up to lockout.

MUSCLES INVOLVED

Primary: Pectoralis major, anterior deltoid

Secondary: Triceps brachii

EXERCISE NOTES

If you find it tough to maintain rigidity in the upper and lower body, push your glutes up slightly into a roughly 10-degree hip angle. This will allow you to easily maintain a rigid position and will also prevent your hips from touching the ground first at the bottom (which limits your range of motion). It will also help maintain high friction between your feet and the floor so that you can get as much stability as possible and maximize pectoral force production. You could also put your feet up against a wall so that friction is no longer a limiting factor.

TIMING CONSIDERATIONS

The deficit push-up is a great exercise if you don't have a lot of equipment or time to wait in line for a machine in a busy gym. It can also be useful when you're in a long phase of fat loss but still want the psychological experience of hitting a personal best. Because you get better at these as you lose weight, deficit push-ups are a great option for the end of a fat loss phase.

SAFETY

Make sure that the blocks or handles you use allow you to go deep enough to achieve a maximum stretch while remaining stable. You do not want your blocks slipping to the sides when you press down. Do not bounce at the bottom, because this can needlessly increase the risk of injury. Most people want to rush the eccentric phase of a push-up in order to eke out more reps, but you should try to avoid this and use a controlled eccentric phase like any other exercise.

VARIATIONS

To enhance the potential of this exercise, consider placing your hands at different widths, from very close together to very far apart. You can keep your legs on the ground or elevate them on a bench or box behind you, which can allow you to target the upper pecs. To increase the loading, you can use a weighted vest or have a spotter place weight on your back, but this comes at a tradeoff with convenience and stability. Ideally, this exercise should be performed without external load (perhaps at the very end of your chest training session if you're very strong), but loaded options can work.

MACHINE PRESS

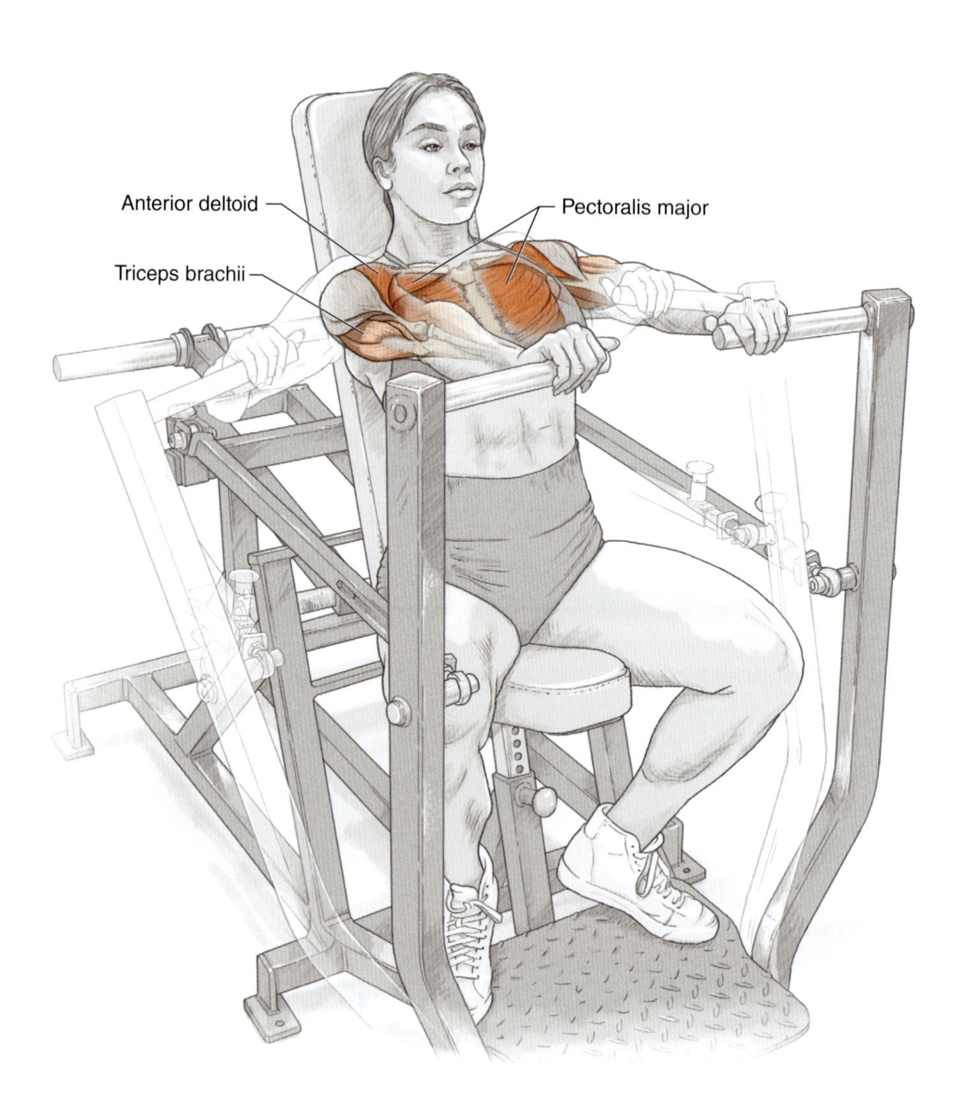

EXECUTION

1. Sit on the chest press machine with your feet on the floor. Adjust the seat so that your arms are horizontal when fully extended.
2. Begin at the bottom of the pressing position. Arch your back, retract your scapulae, and press all the way to lockout.
3. After you achieve a full lockout, come immediately back down slowly, with control. Do not pause at lockout.
4. Pause for a second at the bottom, then press back up. Focus on the deep stretch to the pecs by keeping your chest pushed up throughout the range of motion (especially when closer to the bottom of the press).

MUSCLES INVOLVED

Primary: Pectoralis major, anterior deltoid
Secondary: Triceps brachii

EXERCISE NOTES

Most chest press machines have a variable seat height, so feel free to play with it to find the position that gives you the best stimulus-to-fatigue ratio. Ideally, the seat height should put the handles at roughly your armpit height (or a bit lower or higher based on where you get the most stretch for your pecs with as little shoulder joint discomfort as possible).

TIMING CONSIDERATIONS

Pressing machines are great for workouts that are more time constrained, since setup is minimal and there is no need to rack or unrack weights or position an implement. The machine press is also great for use at the end of a training session, since it requires less technical focus than a free weight exercise.

SAFETY

Make sure to grab the handles evenly so that you don't overly stress one shoulder. Sometimes it is easier to make this mistake with a machine than with a barbell, since the barbell will lean to tell you that your grip is off.

VARIATIONS

Many chest press machines have different handle options. We recommend finding a handle you like and working with it for a few months, progressing in weight or reps during that time. You can continue to use that handle, or try one of the other handles to create variation or alter hand positions. Don't use all of the different handles in the same workout though, because this will use up time with little or no benefit.

DUMBBELL FLYE

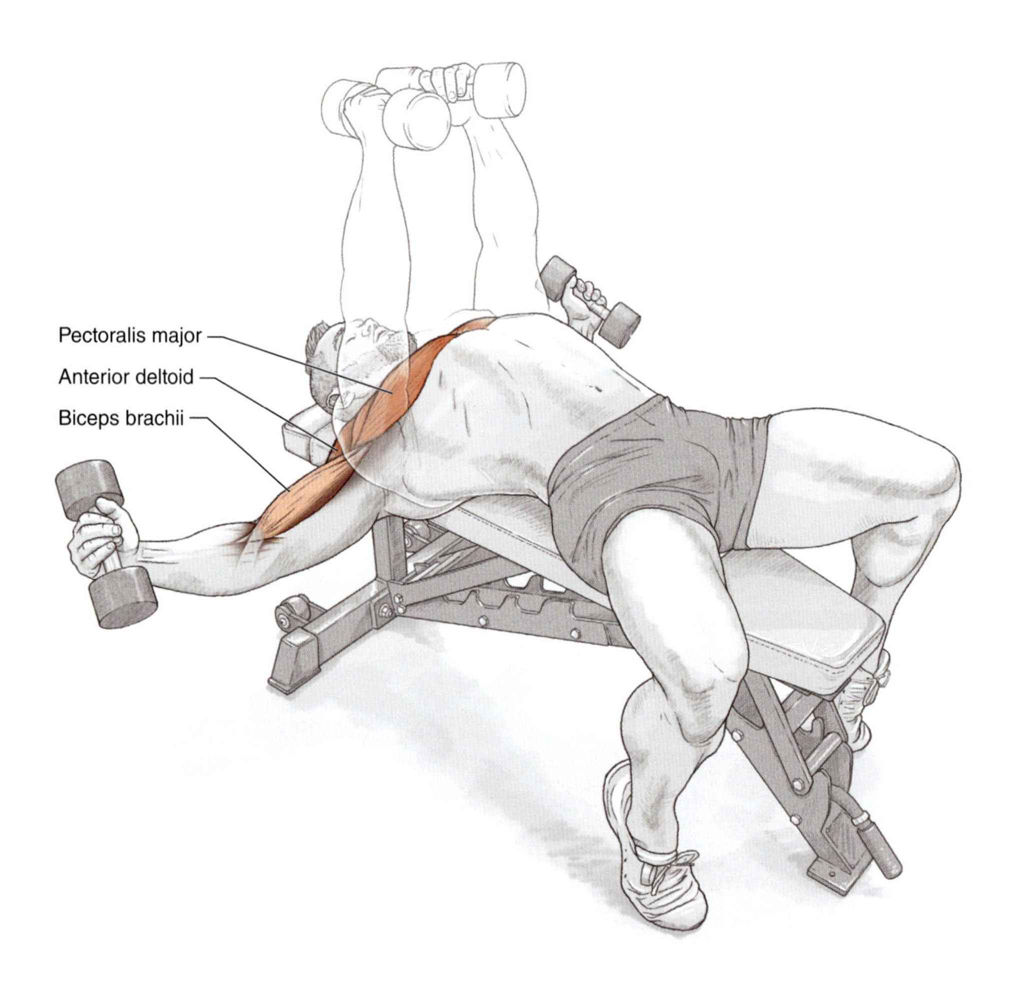

EXECUTION

1. Lie on a flat bench, and extend the dumbbells straight up over your chest in a neutral position (palms facing each other). Unlock your elbows to allow a roughly 15-degree angle of elbow flexion (bend). Keep this angle totally stable throughout the set.
2. Lower your arms to your sides and out, as if you're preparing to hug a big tree.
3. As you lower the dumbbells to chest level and below, reach your chest up and out, stopping for a second in the maximally stretched position and returning to the initial position. You should make an arc with your arms in both the ascent and descent.

MUSCLES INVOLVED

Primary: Pectoralis major, anterior deltoid

Secondary: Biceps brachii

EXERCISE NOTES

In most cases for this exercise, you shouldn't straighten your arms. By keeping them just shy of lockout, you can prevent strain on the shoulder joints and prevent the biceps from stretching so much that they become a limiting factor. In addition, the cue of "chest up at the bottom" can help you both feel the pecs much better and reduce the probability that you experience shoulder joint irritation.

TIMING CONSIDERATIONS

Don't go heavier than eight-pound weights (3.5 kg), unless you're in an intentionally slow eccentric phase. The probability of injury is low but is not totally nominal. The pecs can be finicky, so do not needlessly risk a strain by going ultra heavy.

Most people should only include flyes in their programs if they need isolation work for the pecs. This work can be required if you're hitting the rest of your body so hard that more pressing work would bring too much systemic fatigue or if you're particularly focused on improving your pecs relative to your triceps. Most people can do variations of presses and get all the chest growth they are looking for in less time, since presses allow for triceps growth at the same time.

SAFETY

Quick reversals at the bottom are unwise, because they ramp up the forces on the pecs and their tendons with seemingly no benefit. Gently descend into the bottom position, ideally pause for one second or so, and then gently ease back up. If your shoulder joints hurt when performing this exercise, consider altering your technique and cutting your depth to something that still allows you to feel a deep pec stretch but that doesn't create pain in your shoulder joints.

VARIATIONS

The degree of shoulder adduction will be a comfort issue for most. Technically, you can keep the upper arms flush to your body and still engage the pecs effectively. On the other hand, farther out is better for engaging the pecs but risks creating pain in the shoulder joints. Start with 70 degrees of adduction from the midline and go from there to explore whether small increases or decreases in the angle of adduction create a better stimulus and decrease fatigue.

INCLINE CABLE FLYE

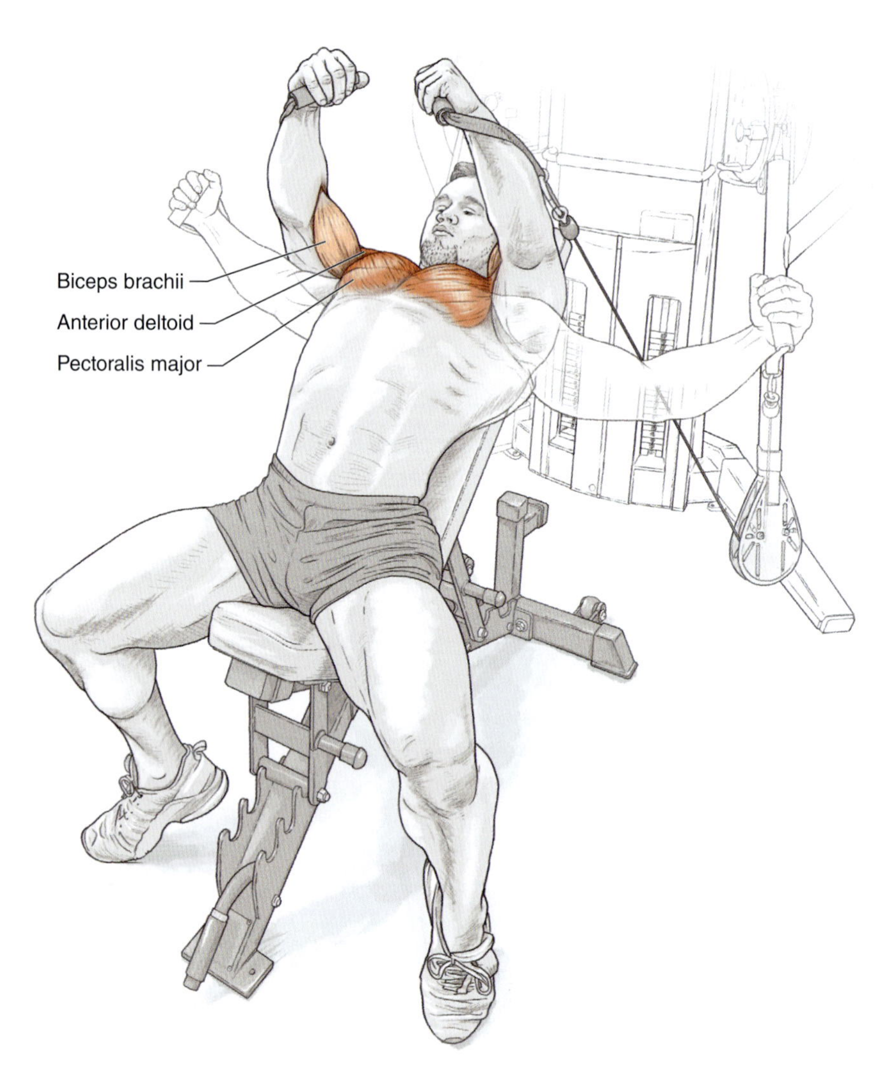

EXECUTION

1. Position and lock two cables of a standard cable tower or freemotion machine in the low position.
2. Whether standing or sitting on a bench, face away from the machine and grab both handles. Remain in this spot throughout the exercise.
3. Perform the same motion you would perform in a dumbbell flye, opening your arms wide to make an arc. Move your arms both up and out so that your upper pecs contribute most of the force.
4. Descend with a 15-degree elbow angle, pause for a second, and flye back up to the starting position, maintaining the same elbow angle on both the ascent and descent. When in the bottom position, keep your chest pointed upward for a deep stretch.

MUSCLES INVOLVED

Primary: Pectoralis major, anterior deltoid
Secondary: Biceps brachii

EXERCISE NOTES

Walk forward as much as you can with the cables before beginning the movement. If you stay close to the stacks, most of the force will be at the peak contraction and not at the stretch. While that's fine for variation, limiting forces at the stretch also limits the most growth-promoting portion of the exercise.

Consider touching your fingers or the handles together for a long one-count at the top of the exercise, to create a peak contraction. This isn't mandatory and may not even enhance muscle growth, but it is a fun variation that can make this exercise feel different from the dumbbell version.

TIMING CONSIDERATIONS

The incline cable flye can be tough to pull off in a busy gym because you need two cable stacks or a freemotion machine all to yourself. Make sure you can have regular, mostly uninterrupted access to the machine if you are going to program it into your next mesocycle

SAFETY

Typically, the incline cable flye is a very safe exercise, but make sure not to use body English. This not only needlessly adds fatigue, but it also makes it difficult to track progress, since your leg strength must now be factored into your chest strength.

You might have to lean forward more as the load gets heavier, which means you may have to reach higher in the concentric phase to keep the cables at an incline. Remain steady throughout. If you are piling on the heaviest weight, this exercise might be nearly impossible to execute, because the load you're trying to move is higher than the frictional forces opposing it between your feet and the ground.

VARIATIONS

If you place an incline bench between the cable towers, you can perform this exercise exactly like you would an incline version of a dumbbell flye. This is a great option, but the setup can be laborious. However, if you are strong enough to need the bench support, or if you have an empty gym and access to custom equipment setups, give it a shot.

DUMBBELL PRESS FLYE

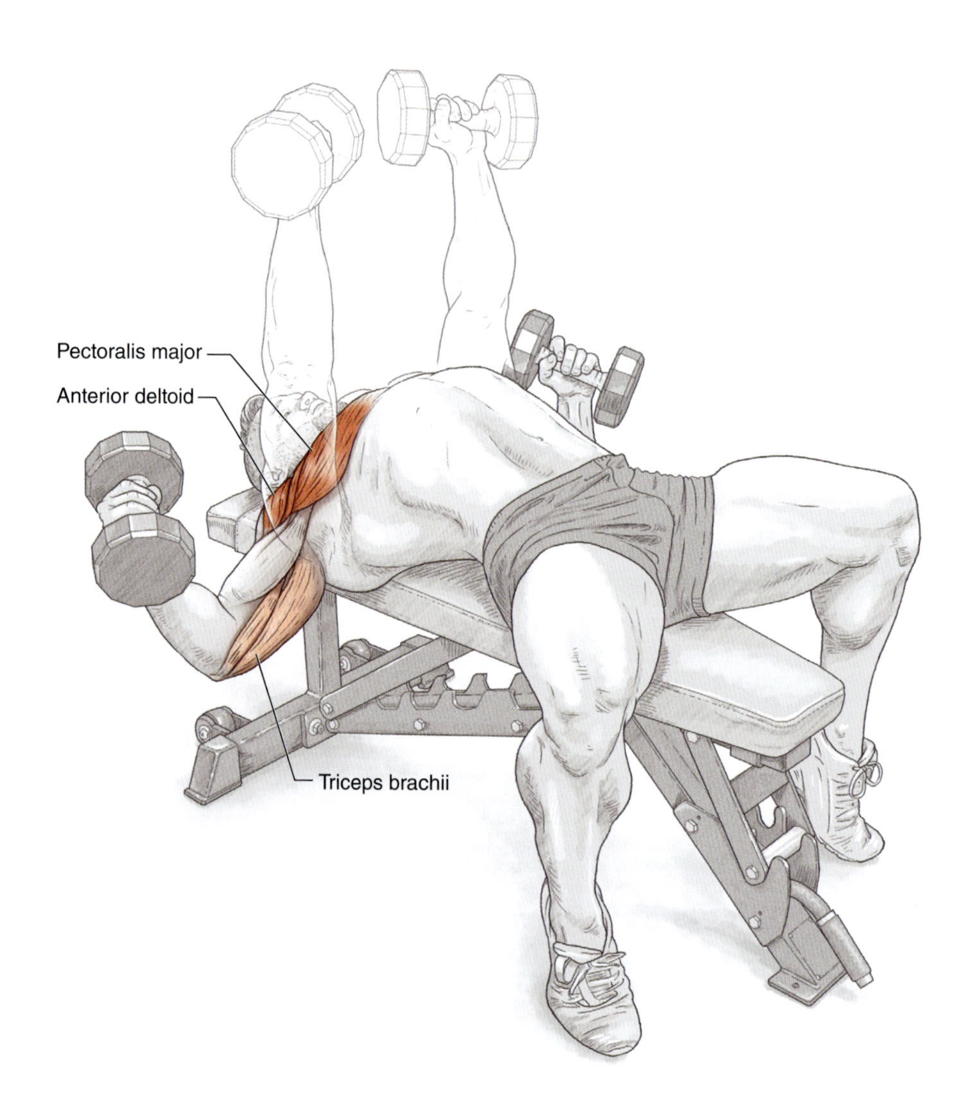

EXECUTION

1. Start by lying supine on the flat bench.
2. Begin the motion at the top, with arms extended vertically in lockout (hands side by side). Keep your feet flat on the floor. If your feet do not reach the floor, you may keep your toes planted on the floor, or you may consider putting a weight plate underneath your feet.
3. Create an arc as you descend, moving the dumbbells out and away from your body. Allow the elbows to bend to about 120 degrees of flexion (or 70 degrees if you measure from full extension). Your elbows should be considerably more bent than they would be in a typical flye, but stop short of going any deeper than just shy of 90 degrees.

4. As you descend into the deep stretch at the bottom, your dumbbells should become parallel to your body. Keep your elbows at 120 degrees of flexion. This is a larger degree of flexion than the 170 degrees or so seen with a flye. Since the 120-degree angle is conserved throughout the bottom range, the exercise is only a press at the top third of the exercise (when the dumbbells are extended from 120 degrees of elbow flexion into 180 degrees and fully locked out).
5. Begin another rep.

MUSCLES INVOLVED

Primary: Pectoralis major, anterior deltoid

Secondary: Triceps brachii

EXERCISE NOTES

It will be tempting to create a deeper angle at the bottom. Allowing more elbow bend than 120 degrees as you descend to the bottom will improve your mechanical advantage by reducing the lever arm of the shoulder joint and allow you to lift more weight—but turns the movement into a press only. If the triceps are brought in by allowing this movement to turn into a press, they will become heavily loaded and stretched at the bottom, creating both growth and fatigue for the triceps. Our goal for the dumbbell press flye is to specifically avoid the involvement of the triceps.

We highly recommend practicing this exercise with very little weight at first, until you have the mechanics figured out. You can then add more weight, but resist the temptation to turn the exercise into a press only. This will take work but will allow the exercise to target the pecs effectively without as much need for external load (saving the joint fatigue you would have accrued) or involvement of the triceps (saving energy for when you want to specifically train the triceps).

As you descend into the 120-degree elbow angle, try to imagine your elbows as completely immovable joints, and reach your chest up as you descend to a deep stretch in your pecs, ideally going as low as you can without losing that 120-degree angle or creating pain in the joints.

TIMING CONSIDERATIONS

This exercise works best if performed by a more experienced trainer rather than a beginner. Before you use this exercise in your training, we highly recommend gaining a lot of experience using flyes and presses by themselves, so that you know the proper executions for both and the differences between them.

> *continued*

Dumbbell Press Flye > *continued*

SAFETY

This exercise can be performed by individuals whose elbows or shoulders feel a bit too strained in the dumbbell flye (especially if the biceps muscle isn't as flexible as the pecs and becomes the limiting factor). Be sure to control the eccentric the whole way through and consider a one-second pause at the bottom to reduce amortization forces for a bit of extra safety. Go only as deep as your shoulder joints are comfortable. You may be able to go deeper over time.

VARIATIONS

This exercise can be performed while lying flat, on a decline, or on an incline. Variations may also involve use of cables instead of dumbbells.

CHAPTER 2

Shoulders

The shoulders are a pivotal muscle group in bodybuilding, contributing to both an impressive upper body physique as well as strength. Well-developed shoulders not only enhance the overall appearance of the upper body, but they also play a critical role in stabilizing a wide range of arm movements. This chapter will provide a comprehensive overview of exercises that effectively target this region without sacrificing shoulder joint stability.

The glenohumeral joint (shoulder joint) is a ball-and-socket joint that allows for a dynamic articulation between the scapula (shoulder blade) and the humerus (upper arm bone) and plays a central role in shoulder function. The glenohumeral joint allows for an extensive range of motion, making the shoulder one of the most mobile joints in the body. However, the trade-off with increased joint mobility is an increase in joint instability, making the shoulder more susceptible to injuries.

Of note, the acromion is a bony extension at the top of the scapula (shoulder blade) that meets the clavicle (collarbone) at the highest point of the shoulder and projects over the glenohumeral joint. The acromion plays an important role in range of motion. There are three common types of acromion, classified according to their shape: type I (flat), type II (curved), and type III (hooked). Type I acromion subtypes have a flat undersurface, resulting in the greatest space between the acromion and the head of the humerus. This extra space allows for a greater range of motion in the shoulder and a decreased risk of shoulder impingement issues, as the rotator cuff tendons pass through this space. Type II acromion subtypes have a slightly curved shape, with less space than type I. If you have a type II acromion, you may notice you have less range of motion in the shoulder, particularly with overhead movements or if the arm is lifted above 90 degrees. Type III acromion subtypes have a downward sloping projection, considerably reducing the space between the acromion and the humerus. If you have a type III acromion, you may find it difficult to perform some of the exercises listed in this chapter, such as the super-ROM dumbbell lateral raise. However, we have included plenty of variations that will target the muscles without irritating the shoulder joint.

The shoulder is composed of several key muscles: the deltoids, the trapezius muscles, and the rotator cuff muscles (supraspinatus, infraspinatus, teres minor, and subscapularis) (figure 2.1). Even though the rotator cuff muscles are not entirely visible compared to other muscles from a purely bodybuilding perspective, it's worth mentioning that the rotator cuff muscles are super important for shoulder stability and strength.

The deltoid is a large, triangular-shaped muscle that forms the rounded contour of the shoulder. It is anatomically divided into three sections, or heads: anterior (front), lateral (middle), and posterior (rear). The anterior deltoid originates from the outer portion of the clavicle and is primarily responsible for moving the arm forward (shoulder flexion). The lateral deltoid originates from the scapula at the acromion and is mostly responsible for moving the arm away from the body in the frontal plane (shoulder abduction). The posterior deltoid originates from the spine of the scapula and is responsible for both extension and horizontal abduction of the shoulder. All three of the heads converge into a single tendon that inserts into the humerus in the upper arm.

The trapezius muscle is a large, flat, diamond-shaped muscle located in the upper back, extending from the base of the skull all the way down to the middle of the back and out to the scapulae. The trapezius is divided into three regions: superior (upper), transverse (middle), and inferior (lower). The superior fibers originate from the base of the skull and neck and insert into the clavicle. They are responsible for elevating the scapula so you can shrug your shoulders during an upright row or lateral raise. The transverse fibers originate from the upper back and insert into the acromion and spine of the scapula. They are primarily responsible for scapular retraction, allowing the shoulders to pull backward. The inferior fibers originate from the middle of the back and insert into the spine of the scapula. They are responsible for depressing the scapula downward.

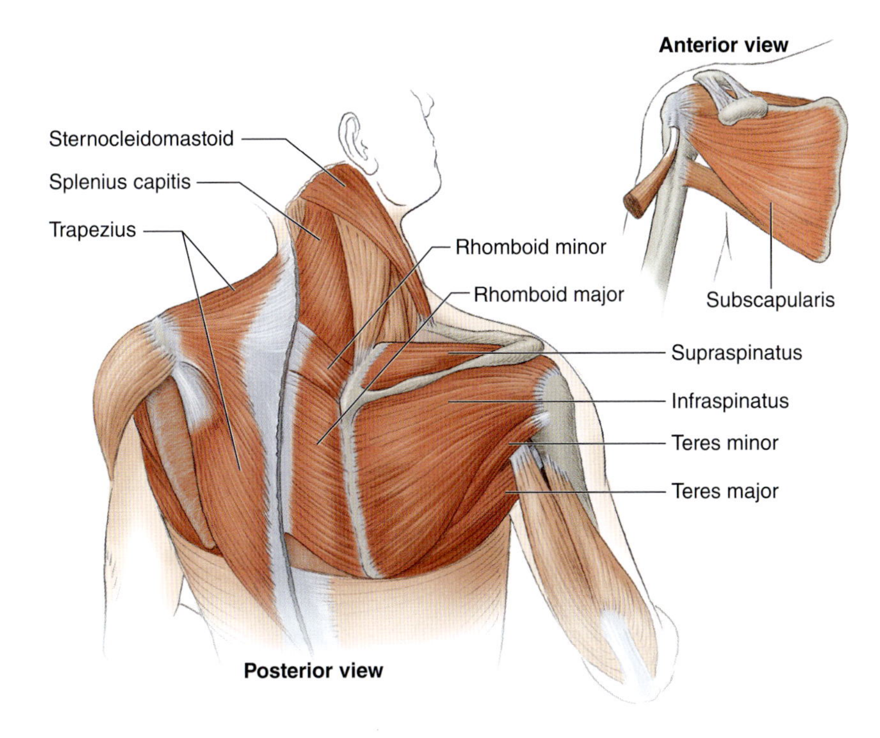

a

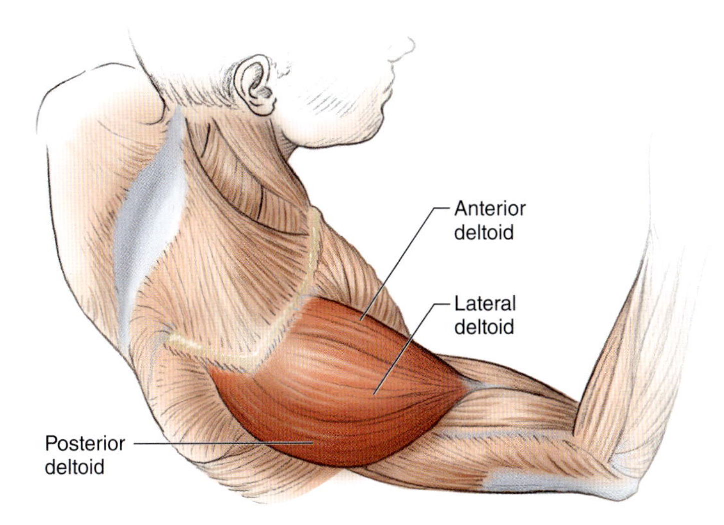

b

Figure 2.1 (*a*) Scapula and rotator cuff; (*b*) deltoid.

DUMBBELL LATERAL RAISE

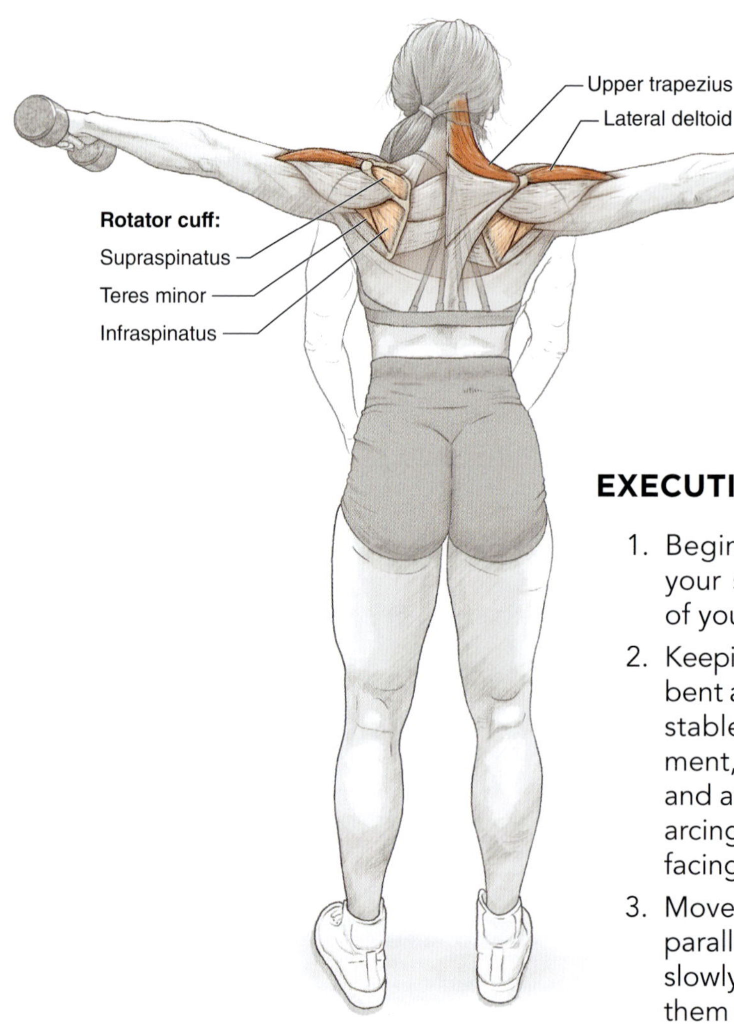

EXECUTION

1. Begin with the dumbbells at your sides or just out in front of your hips.
2. Keeping the elbows only slightly bent at around 10 degrees and stable throughout the movement, raise the dumbbells up and away from your sides in an arcing motion, with your palms facing down the whole time.
3. Move your arms to just above parallel with the ground. Then slowly and with control, return them back to your sides.

MUSCLES INVOLVED

Primary: Lateral deltoid, upper trapezius

Secondary: Rotator cuff (infraspinatus, supraspinatus, subscapularis, teres minor)

EXERCISE NOTES

If you keep your palms down, you mostly engage the side delt (lateral aspect of the deltoid muscle), even if you aren't pulling your arms directly up to the sides. You can go as much as 45 degrees toward the sagittal plane (the front of the body) and all the way back to 15 degrees behind the frontal plane with great side delt stimulus. Try a few different motion paths, and see which one gives you a good combination of side delt stimulus without irritating the shoulder joint. Your best bet for a robust stimulus is to move the dumbbells up to at least 15 degrees above parallel to the ground, so a few tracks to that position might work better than others.

The higher you can tilt your pinkie fingers (and the lower you can point your thumbs), the less you will involve the anterior deltoid and the more you will involve the lateral and posterior deltoid. However, the more you do this "pouring out a glass of water" tilt with your hands, the more likely you are to encounter shoulder joint discomfort at higher ranges of motion. While you should use this positioning, prepare to make a trade-off so that you go the highest you can with your pinkie fingers up but where your shoulder joints feel best. There could be many "correct" ways to do the movement, so just pick one technique and be consistent with it for a few months of training (potentially altering it later or just sticking to it indefinitely).

Similar advice applies to the elbow angle. While most people experience shoulder or elbow joint irritation from locking out their elbows, some prefer it, and some prefer a very large elbow bend to a very small one. Whatever bend you pick that feels best for your joints, stick to it for several months before switching, and never choose it simply because it lets you lift more weight. You can always do an upright row if you'd like to lift heavier dumbbells!

TIMING CONSIDERATIONS

The lateral raise is ideal for use in supersets with dumbbell upright rows. Right after you finish a set of lateral raises, switch to dumbbell upright rows with the same dumbbells. This will really burn your side delts in the best way possible.

SAFETY

If you descend very slowly with the dumbbells (taking three to five seconds on the eccentric phase), this maximizes the growth stimulus while reducing strain on your connective tissues. This might be an impossible task if you're swinging the dumbbells up to get to the top, so stop doing that! Controlled reps create controlled outcomes, and those just happen to be the best long-term muscle growth and joint health outcomes as well.

VARIATIONS

If you're looking to spice things up, try a few months of doing laterals with a full one-count pause at the very top, followed by the usual slow eccentric. You can also try these with a roughly 15-degree forward lean to your torso. You'll need to use less weight and won't be able to raise your arms as high up, so the stimulus to the lateral and posterior aspects of the delt can be intense.

SUPER-ROM DUMBBELL LATERAL RAISE

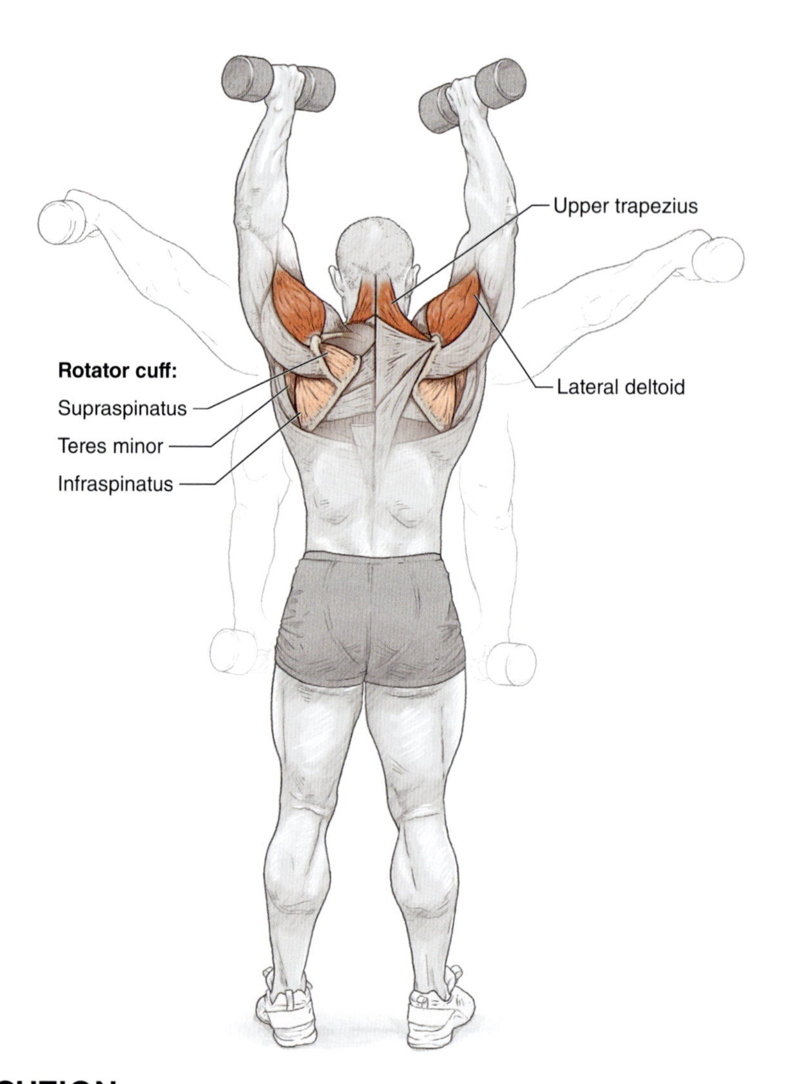

EXECUTION

1. Begin with the dumbbells at your sides or just out in front of your hips.
2. Keeping the elbows only slightly bent at around 10 degrees and stable throughout the movement, raise the dumbbells up and away from your sides in an arcing motion, with your palms facing down the whole time.
3. Move your arms as high overhead as they can go. Then, slowly and with control, return them back to your sides.

MUSCLES INVOLVED

Primary: Lateral deltoid, upper trapezius muscle

Secondary: Rotator cuff (infraspinatus, supraspinatus, subscapularis, teres minor)

EXERCISE NOTES

This movement is easiest and most natural to perform if you keep your pinkie fingers down and thumbs up the whole time—but then it also engages mostly the anterior aspect of the delt and not the lateral aspect. To best engage the lateral deltoid, keep your pinkie fingers as high up as you can until you're above parallel, and then turn your thumbs up just enough to make the top portion comfortable on your shoulder joints. Once you've reached the top, slowly turn your pinkie fingers back up and your thumbs down as you descend. If this sounds complicated, don't worry—it will begin to feel natural very soon.

At the top you can touch your palms together, but you don't have to. As long as your arms are high enough to no longer feel gravity pulling on the dumbbells much, you can reverse at any time.

When you get to the top and reverse, don't rush back down. Engage your side delts right away to slowly grind the eccentric all the way. It's a long way down, so buckle up for an intense deltoid growth stimulus. It's especially important to engage and maintain eccentric control through the middle 90 degrees of the motion (from 45 degrees above parallel to 45 degrees below it) where the forces are highest.

TIMING CONSIDERATIONS

Because this movement is a bit more complex, it probably makes most sense to master the dumbbell lateral raise before experimenting with the super-ROM dumbbell lateral raise.

SAFETY

If you experience consistent shoulder joint pain when performing this exercise, reduce the top-end ROM, stop doing the top-end ROM, or stop the exercise completely for a while. There's nothing magical about this exercise that you can't get out of other exercises, and some people's shoulder joints are just a bit more irritable than others.

VARIATIONS

If you have access to an Atlantis lateral raise machine, you can enjoy the benefits of this exercise and then some. Their machine applies roughly even tension through the whole movement, even at the bottom and top where gravity doesn't resist much in the dumbbell version. With both this machine and dumbbells, you can try a variation for a few months where you raise your arms to only 45 degrees above parallel to the ground and 45 degrees below parallel to the ground, keeping constant high tension on the muscle. This isn't a superior variation, but it is effective.

FREEMOTION Y-LATERAL RAISE

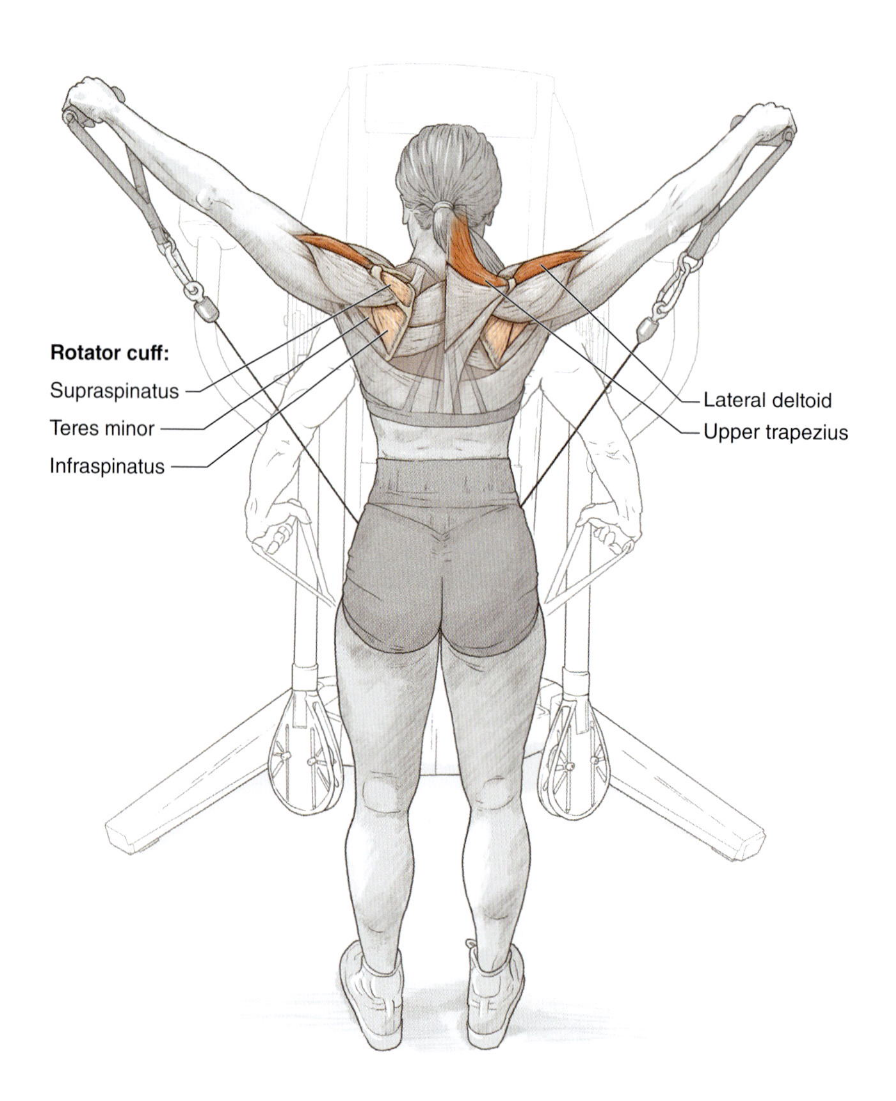

EXECUTION

1. Position the cables in the low position. Begin holding the cable handles at your sides or just out in front of your hips.
2. Keeping the elbows only slightly bent at around 10 degrees and stable throughout the movement, raise the cables up and away from your sides in an arcing motion, with your palms facing down the whole time.
3. Move your arms as high up overhead as they can go. Then, slowly and with control, return them to your sides.

MUSCLES INVOLVED

Primary: Lateral deltoid, upper trapezius muscle

Secondary: Rotator cuff (infraspinatus, supraspinatus, subscapularis, teres minor)

EXERCISE NOTES

Try to feel your side and rear delts working by finding arm positions that give you the best mind–muscle connection. This can be anywhere between a raise all the way to the side in the frontal plane or a raise almost all the way to the front in the sagittal plane. In most cases, somewhere between the two extremes will feel best for muscles and joints.

The cables allow continual high tension through the entire range of motion. Don't miss out on this! Make sure to milk out a slow eccentric on every rep. This allows you to use less weight and gain more muscle growth. The only price is the acute pain of your side delts getting burned with their own metabolites.

TIMING CONSIDERATIONS

This exercise can offer a great break from dumbbell-only shoulder work, which can become monotonous if that's all you do for long enough. This exercise is done rarely enough in the gym that we suspect most people don't know about it!

SAFETY

Don't start bending the elbows to get extra reps or rest-pausing at the top within each rep to get a lockout. With cables, you can do the "grind-pause, grind-pause" dance if you want, but this approach can take a huge psychological toll without as much of a growth stimulus benefit. It also sets you up in a very difficult situation from which to track performance. Keep all reps smooth, and if you momentarily stop midway through the concentric of a final rep, give yourself only one chance to keep it moving again (not multiple chances within the same set or even the same rep).

VARIATIONS

Where you position the cables and which way you arc your hands are huge factors in creating variations. You can bias different sides of your deltoids and make the movement feel better or worse by adjusting cable positioning and how far you step away from the cables. Try a few variants and choose the ones that feel best, staying open to the idea that you might land on some even better variants later.

EZ BAR UPRIGHT ROW

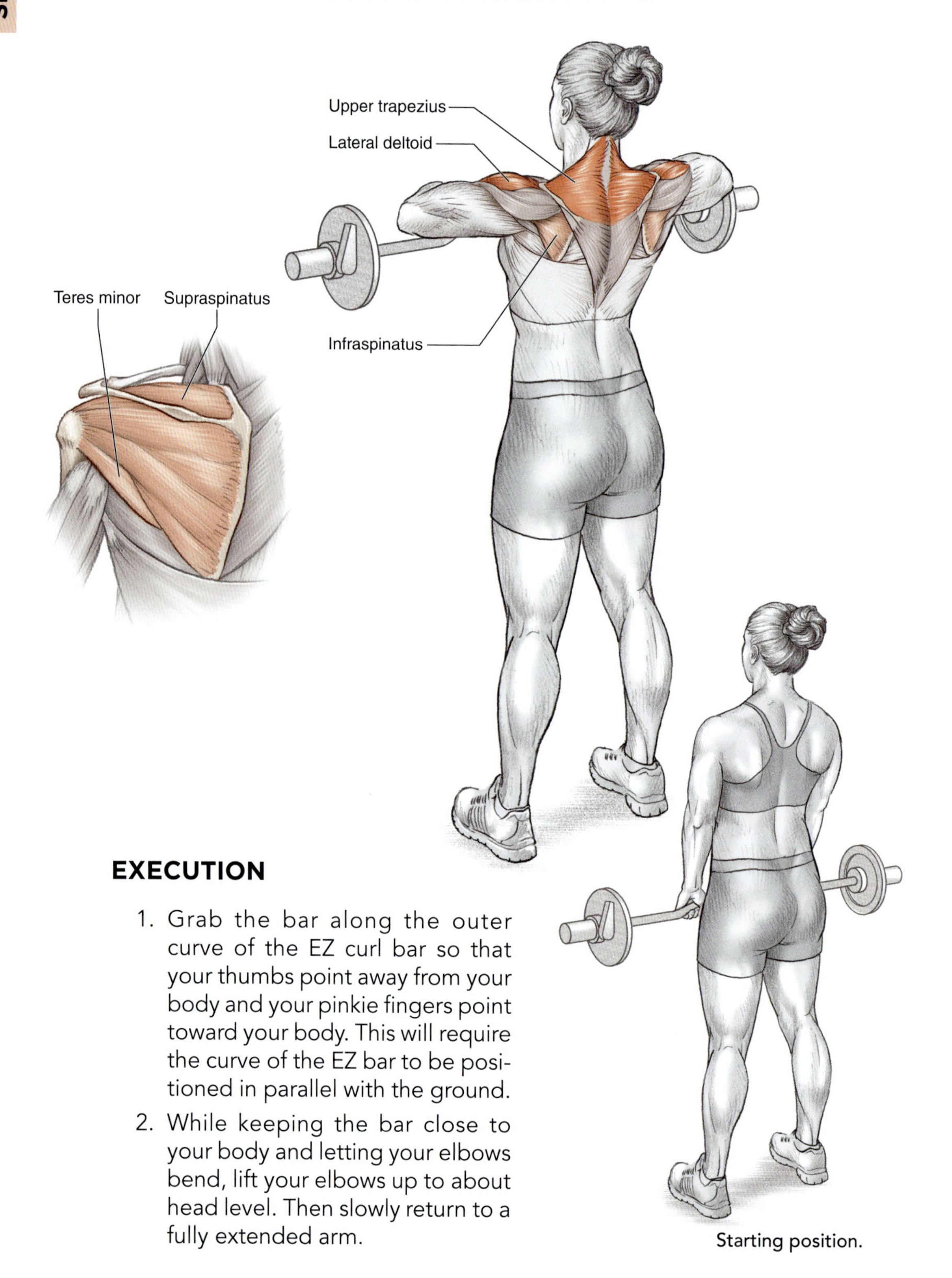

Starting position.

EXECUTION

1. Grab the bar along the outer curve of the EZ curl bar so that your thumbs point away from your body and your pinkie fingers point toward your body. This will require the curve of the EZ bar to be positioned in parallel with the ground.
2. While keeping the bar close to your body and letting your elbows bend, lift your elbows up to about head level. Then slowly return to a fully extended arm.

MUSCLES INVOLVED

Primary: Lateral deltoid, upper trapezius

Secondary: Rotator cuff (infraspinatus, supraspinatus, subscapularis, teres minor), biceps brachii, brachioradialis, brachialis

EXERCISE NOTES

An important cue in the upright row is to lead with your elbows. This can help you feel out and activate the lateral delts more, instead of leading with your hands and using the forearm flexors more in a curling-style motion. In addition, as your elbows pass chest level on the way up, pull them backwards as well as up. This can enhance the lateral delt mind–muscle connection and create an extra bit of posterior aspect activation.

TIMING CONSIDERATIONS

While this exercise is never a bad choice by itself, it's also an option to try if your wrists are not fond of straight bar upright rows.

SAFETY

Not everyone has the shoulder structure to prefer or tolerate the same hand position and top-end range of motion on upright rows. If these hurt your shoulder joints, consider narrowing or widening your grip. If that doesn't resolve the issue, reduce your range of motion at the top until you find the largest ROM that doesn't irritate your shoulders.

VARIATIONS

There are numerous ways to vary this exercise. First, you can vary your grip width from the outer curves of the EZ bar to the inner curves, or you can even grab the bends between them. Grip preference is often very personal in this exercise, with some people swearing by a closer grip, some swearing by a wider grip, and some who can use both variations for several months at a time.

Second, you can vary the degree to which you let the barbell float forward as you pull up from the first half to the second half of the pull. Some individuals achieve a better mind–muscle connection from pulling forward a bit at the bottom and then pulling back at the top, while others prefer pulling in a straight line close to the body.

Lastly, as long as you're pulling to at least your nipple line, there's no right or wrong pulling height. You can stop the pull anywhere between the nipple line and above your head, depending on where you feel it the best or even just for variation every few months.

ONE-ARM CABLE REAR DELT FLYE

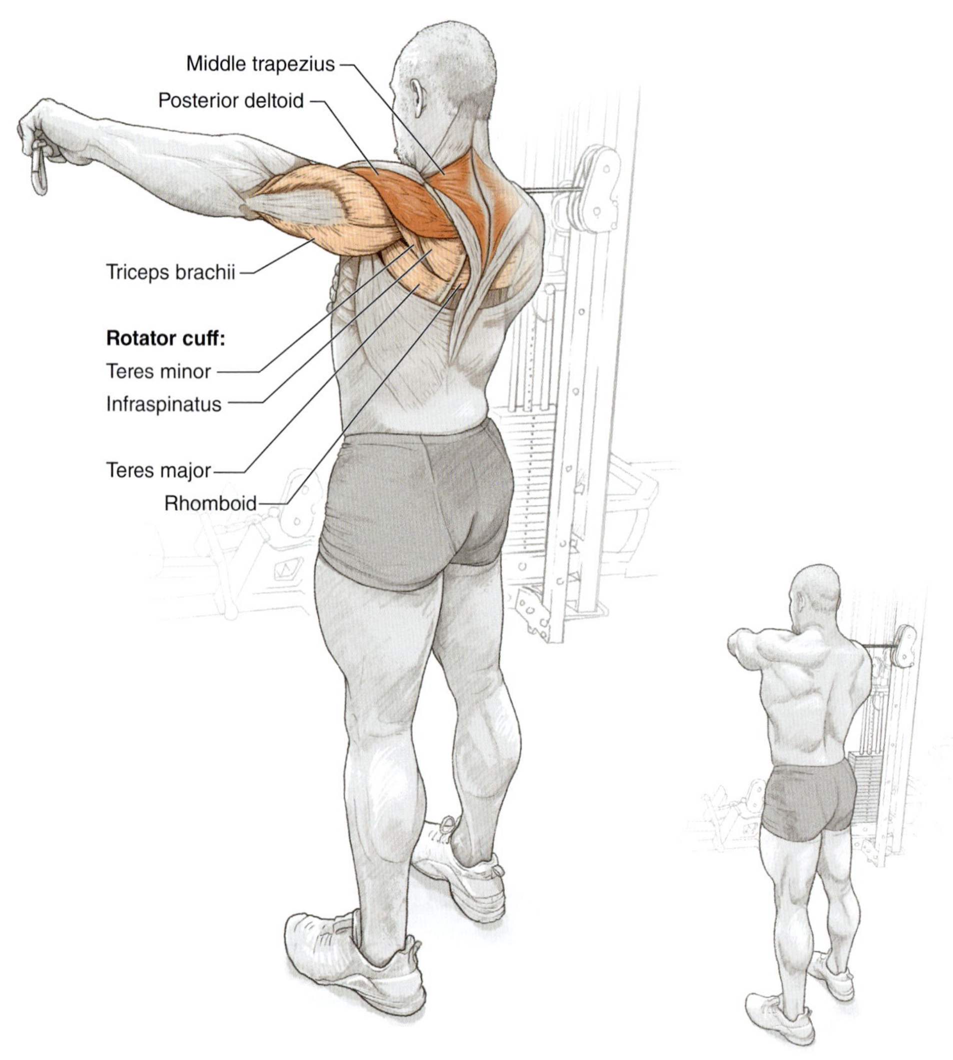

Starting position.

EXECUTION

1. Set the cable stack height to anywhere between hip and shoulder level.
2. Grab the cable with one hand and walk away from the stack. Keep your opposite shoulder close to the stack so the cable runs just across your body.
3. Keeping your elbow locked out or close, pull your arm across your body in an arcing motion until your arm is pointed in the opposite direction of the cable. Slowly and with control, return to the starting position.
4. Switch sides after completion of the entire set.

MUSCLES INVOLVED

Primary: Posterior deltoid, middle trapezius

Secondary: Rotator cuff (infraspinatus, supraspinatus, subscapularis, teres minor), teres major, rhomboid, triceps brachii

EXERCISE NOTES

Make sure that your body is perpendicular to the cable stack and your arm is wrapped around your abdomen or chest so that your posterior delt is stretched as much as possible at the beginning. You may be tempted to lean in to leverage your pull better, but leaning away exposes the posterior delt to more tension, which is the main goal of the exercise.

When bringing your arm across, think of an arcing motion that puts your arm as far out from your body as possible (imagine trying to spill a can of paint onto the floor as far away from your body as you can). The temptation will be to bend your elbow and retract your scapula so that the magnitude of the arc decreases; while that allows you lift more weight, it also reduces the mind–muscle connection you have with the posterior deltoid. On the way back down, again try to traverse the biggest arc that you can, while protracting the scapula and reaching your arm forward. Eccentric control is also important here, because rushing the descent will reduce the effectiveness of the exercise (in particular, you'll miss out on a lot of the mind–muscle connection).

TIMING CONSIDERATIONS

Because this exercise requires a lot of space inside a cable stack, it can be tough to set up in a busy gym. It also takes a long time to complete this exercise because it involves using one arm at a time. If you're in a rush in a busy gym, this is not an ideal exercise.

SAFETY

The angle of pull has a lot of impact on how your shoulder joints will feel. If your shoulder joints don't feel good using a low angle (cable height beginning at hips), try a higher angle (cable height beginning at just above shoulder level) and see where your shoulder joint feels best.

VARIATIONS

If you're ok with managing cable snag and making a slightly different motion with one arm than the other in every rep, doing these bilaterally (at the same time with two cable stacks) is a great option. Every other set, alternate which hand goes high and which one goes low.

CABLE FACE-PULL

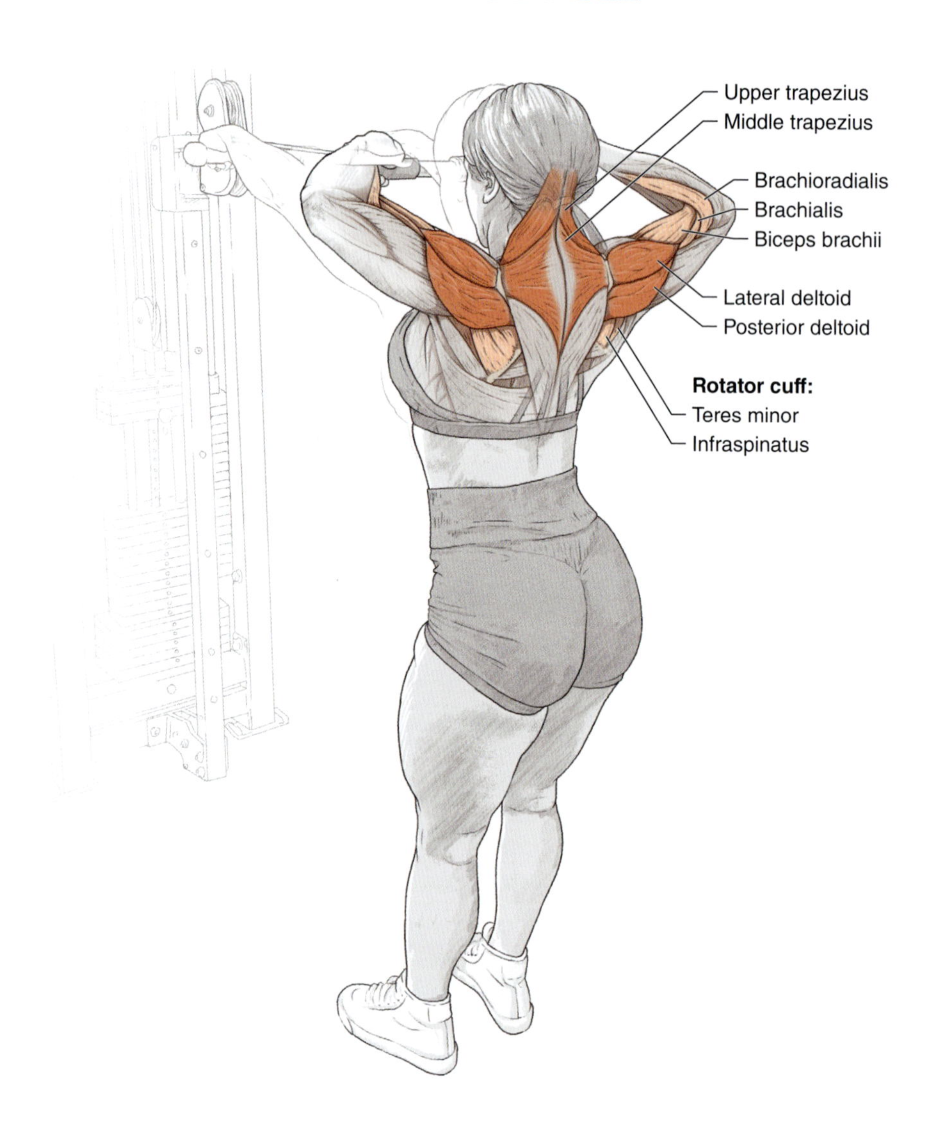

EXECUTION

1. Setting the cable height anywhere between your hips and your head, walk backward with your hands on the ends of the cable attachment until you've taken the slack out of the cable and are exposed to full tension.
2. Bring your elbows back and up until they are as far behind your head as they can go. Stop for a second. Then slowly return the elbows back to full extension.

MUSCLES INVOLVED

Primary: Lateral and posterior deltoid, upper and middle trapezius

Secondary: Rotator cuff (infraspinatus, supraspinatus, subscapularis, teres minor), biceps brachii, brachioradialis, brachialis

EXERCISE NOTES

You can vary the pulling height depending on your delt focus, whether lateral or posterior, or based on what gels with your shoulder joint comfort the most. There is no right or wrong pulling height so long as you feel the aspect of the delt you want to target and your shoulder joints aren't unduly irritated.

Pausing at the top is highly recommended in this exercise. Pausing will allow you to begin the eccentric with a very high mind–muscle connection to your delts and make sure the delts are maximally active all the way through the growth-promoting eccentric phase. This isn't mandatory, but it's a good idea for most to try.

TIMING CONSIDERATIONS

It would be easy to go through the motions on this exercise, using a lot of forearm flexor and trapezius force to complete the movement. If you're choosing to do this movement, make sure you're ready to clue in and curate a high mind–muscle connection on every rep. If you're a beginner and don't know how to do this or you're in a rush and just want a quick all-around workout, there are better shoulder movements to choose from in most cases (such as upright rows).

SAFETY

Never repeat motions that hurt your shoulder joints. If this exercise is hurting them, try to find a different pulling angle or degree of elbow flare that stops hurting them.

VARIATIONS

You can take either an overhand or underhand grip on the straight bar, as both seem to target the delts in subtly different ways. You can also use a rope attachment instead of a straight bar. If you like just one grip, you can use only that one, but if you like both, doing one for a few months at a time and then switching to another is a great option.

MACHINE REAR DELT FLYE

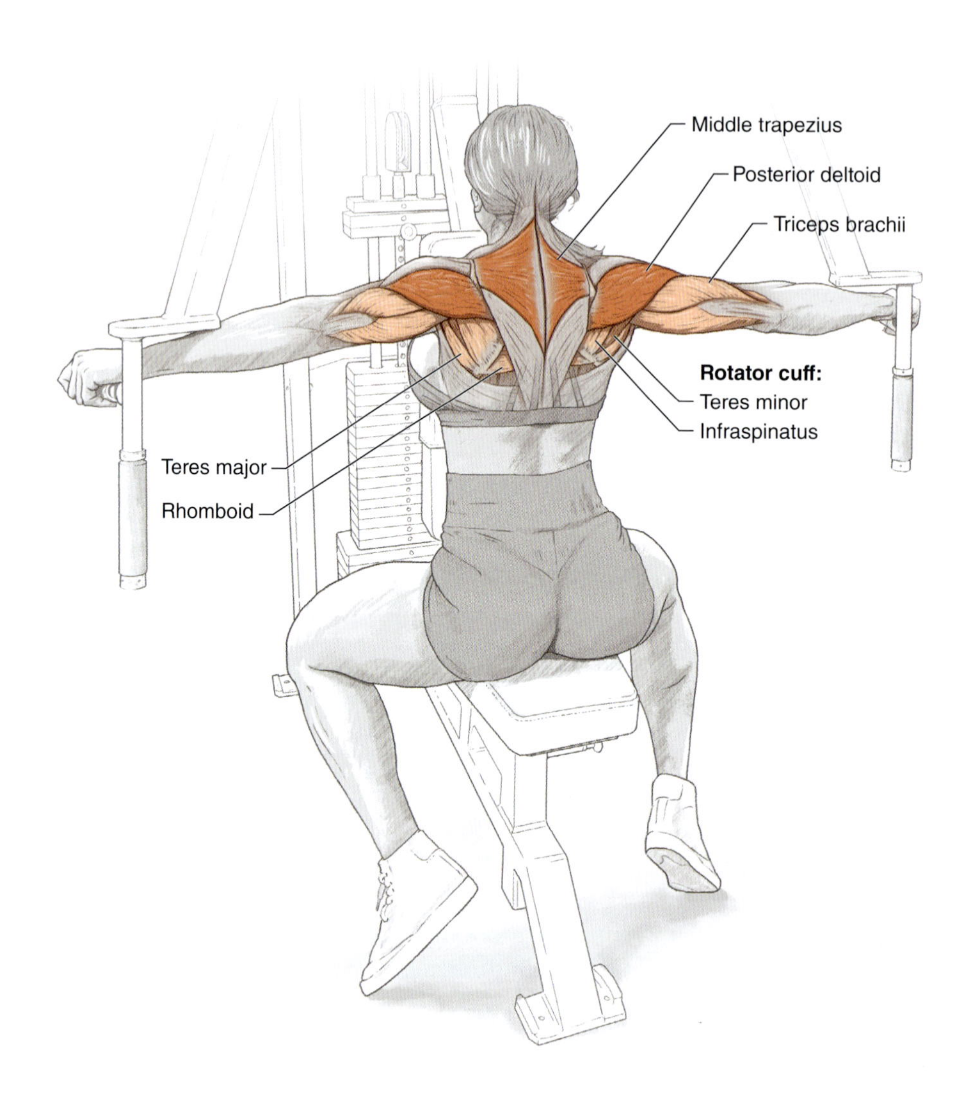

EXECUTION

1. Set the chest support of the machine so that you're reaching very far forward (with protracted scapulae) to grab the handles. Set the seat height so that your arms are parallel to the floor.
2. Arc your arms out and away from you until they are behind you. Then slowly arc them back to the starting position. There should be minimal or no elbow bend during this whole exercise.

MUSCLES INVOLVED

Primary: Posterior deltoid, middle trapezius

Secondary: Rotator cuff (infraspinatus, supraspinatus, subscapularis, teres minor), teres major, rhomboid, triceps brachii

EXERCISE NOTES

Protract your scapulae at the start of the exercise so that you can increase the lever arm of the exercise and make it harder on your rear delts. This may also help enhance the mind–muscle connection with the rear delts. You may be tempted to pull your scapulae back (since it lets you lift more weight), but this will cause your back musculature to feel more of the movement instead of your rear delts. To further assist mind–muscle connection, think of trying to paint as large an arc as you can with the handles. This will reinforce the protracted scapulae as well as the elbow lockout, both of which can help enhance the mind–muscle connection. Lastly, it might help the mind–muscle connection to the rear delts if you pause at the top end of the movement and actively try to contract the posterior delts as you begin the eccentric.

TIMING CONSIDERATIONS

Do this exercise first or early in your session if you need extra rear delt work. Don't do it last because that will lower its stimulative effects. You may also find you do not need a whole lot of extra rear delt work if you're training your back properly—just as a lot of front delt work is not required if you're training your chest properly.

SAFETY

Too much load on this exercise can irritate your shoulders. Focus on technique; mind–muscle connection; and a smooth, slow eccentric phase to maximize the effect while reducing the need for high external loads.

VARIATION

Feel free to play with the seat height to see which position best engages the rear delts and is least irritating for your shoulder joints. If multiple seat heights feel equally great in different ways, feel free to use one of them for a few months, and then switch to another.

SEATED BARBELL PRESS

EXECUTION

1. Unrack the barbell. With your elbows in lockout, let the barbell float directly over your glenohumeral joint.
2. Unlock your elbows, and bring the barbell down to touch your clavicles while keeping it as close to your face as possible during the movement.
3. As the bar descends, lift your chest to meet it. Once it touches your clavicles, push the bar back up.
4. Once the bar clears your face, begin to move it slightly back as well as up until you reach lockout.
5. Repeat the motion.

MUSCLES INVOLVED

Primary: Anterior deltoid, triceps brachii
Secondary: Upper (clavicular) pectoralis major

EXERCISE NOTES

Keep the bar as close to your face as possible on the descent and ascent. This will standardize the movement and allow you to plan overloads and track progression. If the bar floats forward a lot, the leverage for your delts will drop dramatically; it may become nearly impossible to match last week's performance. As you get stronger, you can begin to pause every repetition once the barbell touches the clavicles. This can prolong the loaded stretch on the delts (driving more growth stimulus) and keep the movement safer.

TIMING CONSIDERATIONS

Most people who train their chests intelligently and hard won't need extra front delt work. If you do, this is a great option to do early in a session when you're fresh and strong.

SAFETY

Unrack and re-rack carefully and stably. You don't want to hurt yourself by having the rack position too far back or by hitting the rack on the way up if it's too close. Play with the rack position to get it as workable as possible. Ideally, have a spotter help you unrack and re-rack the bar.

VARIATIONS

You can do presses seated on a machine or with dumbbells. You can also do presses standing with a barbell, dumbbells, or on a machine. There's no shortage of shoulder pressing variations, especially if you play with different grip widths. It does help to remember that standing press options will be more axially fatiguing than seated pressing options. That is something to consider if you have a program that already asks a lot of your spinal support musculature (such as one loaded with lots of deadlifts, good mornings, and rows).

CHAPTER 3

Back

The back, also known as the rear torso, is another major muscle group that is essential in bodybuilding. A well-developed and conditioned back is crucial for bodybuilders who are aiming to optimize their overall physique and performance. A broad, widely developed back creates the look of a smaller waist, creating a V-shaped body that is highly sought after in bodybuilding. The muscles of the back (figure 3.1) can be broadly divided into superficial and deep muscle groups.

The superficial group—which includes the rhomboids (major and minor), trapezius, and latissimus dorsi—is primarily responsible for the movement and stabilization of the shoulder girdle (the clavicle and scapula). The rhomboid major is a rectangular-shaped muscle that originates from the spinous processes of the upper thoracic vertebrae (T2 through T5) and inserts on the medial border of the scapula, from the level of the scapular spine down to the inferior angle. The rhomboid minor is a smaller, rectangular-shaped muscle that sits above the rhomboid major. It originates from the spinous processes of the lower cervical (C7) and upper thoracic vertebrae (T1), and inserts on the medial border of the scapula at the level of the scapular spine. Together, the rhomboids retract the scapula toward the spine and rotate the scapula down.

The trapezius muscle (also known as the "trap") is a large, flat, diamond-shaped muscle located in the upper back, extending from the base of the skull down to the middle of the back and out to the scapulae. This muscle is divided into three different regions: superior (upper), transverse (middle), and inferior (lower). The superior fibers originate from the base of the skull and neck and insert into the clavicle. They are responsible for elevating the scapulae so you can shrug your shoulders. The transverse fibers

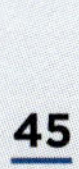

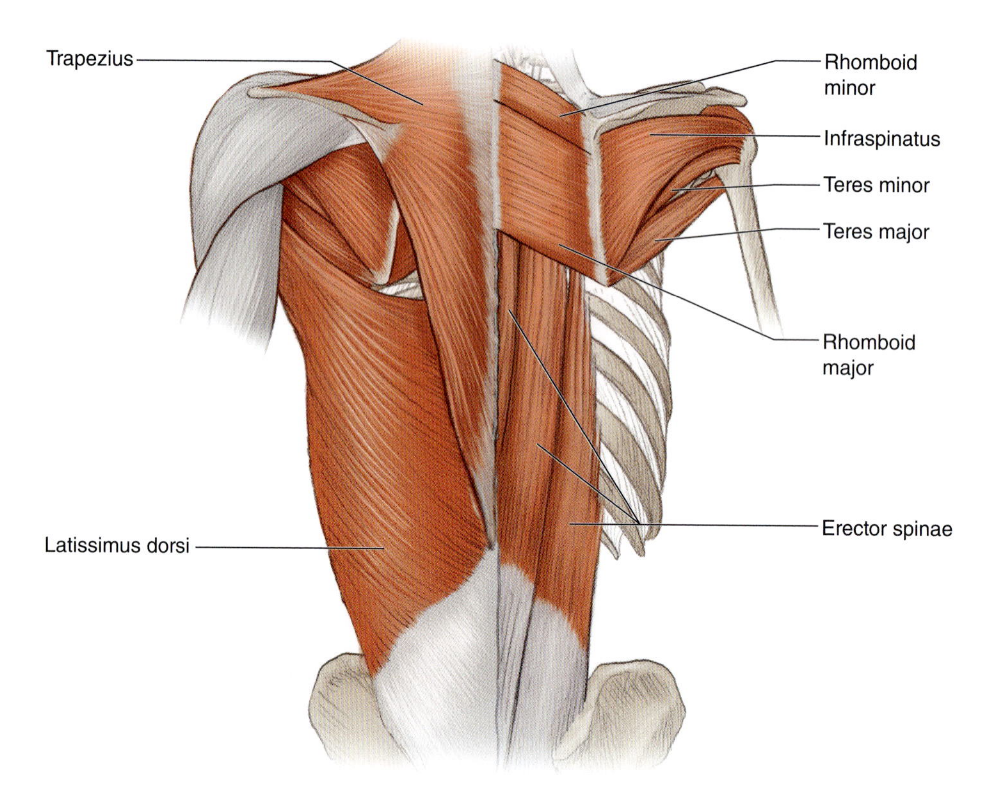

Figure 3.1 Back.

originate from the upper back and insert into the acromion and spine of the scapula. They are primarily responsible for scapular retraction, allowing the shoulders to pull back during most, if not all, rowing exercises. Finally, the inferior fibers originate from the middle of the back and insert into the spine of the scapula. They are responsible for depression and upward rotation of the scapula. Many of the shoulder exercises described in the previous chapter also work the trapezius muscle, mainly the upper region. The exercises in this chapter have a greater emphasis on the middle and lower regions of the trapezius muscle for maximal development.

The latissimus dorsi (also known as the "lat") is a large, fan-shaped muscle that originates from multiple sites: the lower half of the thoracic vertebrae (T7 through T12), the thoracolumbar fascia, and the iliac crest of the hip bones. From its large origin, the muscle fibers converge into a thick tendon that inserts into the humerus, near the pectoralis major. Not surprisingly, the lats are responsible for a few essential movements of the arm: shoulder extension, shoulder adduction, and shoulder internal rotation. Pull-downs, pull-ups, and rows are all great examples of exercises that effectively target the lats.

The deep muscle group in the back includes the erector spinae (iliocostalis, longissimus, and spinalis); the multifidus; and many other smaller muscles. Together, this muscle group plays a vital role in supporting the spine, controlling spinal movements, and maintaining postural alignment. While flexion rows can specifically target and torch these smaller muscles, you'll notice that many of the exercises in other chapters include the erector spinae as a static stabilizer. Keep this in mind when designing your program.

OVERHAND PULL-UP

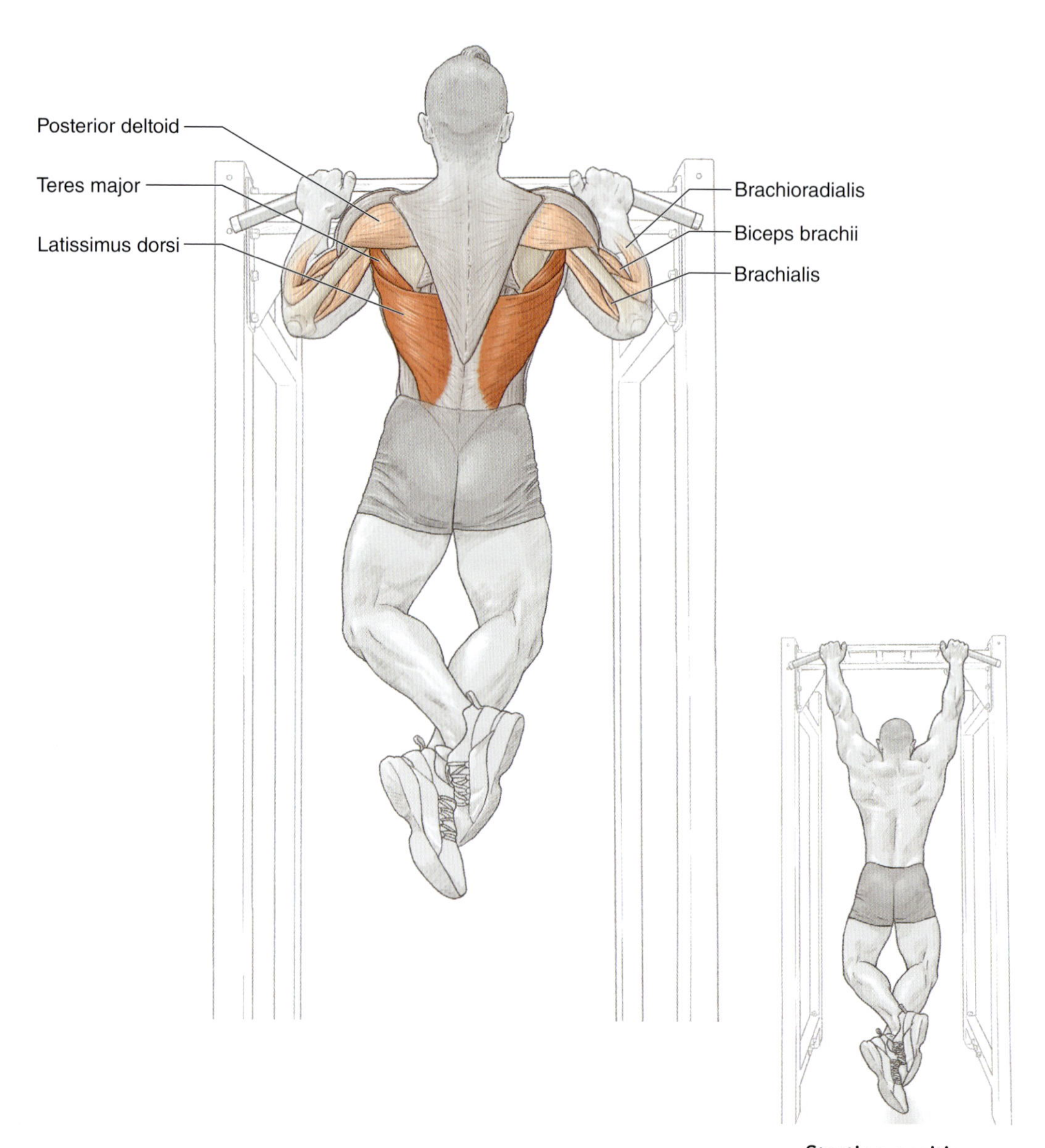

Starting position.

EXECUTION

1. With an overhand grip on the bar and hands about shoulder-width apart, hang down from the pull-up bar. Your arms should be fully outstretched, and your scapulae should be fully relaxed and elevated.
2. Without pulsing too hard on the ascent, gradually accelerate your body up until your chin elevates above the bar.
3. As you pull up, try to pull your elbows down close to your sides toward your outer hips.
4. Control the slow descent all the way back to the starting position.

MUSCLES INVOLVED

Primary: Latissimus dorsi, teres major

Secondary: Posterior deltoid, biceps brachii, brachialis, brachioradialis

EXERCISE NOTES

A looser grip and looser elbows during the descent will maximize the lat stimulus. However, other grips and positions are fine for variation. People rarely control the eccentric phase on pull-ups because it makes them much harder, but it also makes pull-ups more effective. Don't skimp on the deep stretch and dead hang at the bottom. Whether you choose to keep the muscles active during the momentary dead hang is much less important than whether you're getting into that position at all. So, dead hang or active hang, get down there!

TIMING CONSIDERATIONS

If you're not strong enough to do at least five reps when fresh, consider assisted pull-ups and pull-downs for a while until you either build up more pulling strength or are lighter and can do more reps of pull-ups.

SAFETY

Do not bounce at the bottom, no matter how many extra reps it gets you. If you'd like to increase your chances of injury for no upside, by all means, bounce away.

VARIATIONS

A benefit of pull-ups is the variety of grip positions that are possible. These include neutral grip (which involves using parallel handles that face each other) and underhand grip (which involves a supinated grip where the palms face toward you). These grips target different muscles: the neutral grip primarily works the brachialis and brachioradialis in the arms, and the underhand grip targets the biceps. In addition to the grip positions, the width of the grip can also be adjusted to target specific areas of the back. Pull-up variations can also be done with different ranges of motion, such as pulling up until the sternum touches the bar for the most advanced lifters, or just to the chin, clavicles, or even eye level for those who are not quite as strong. These variations can provide a more complete workout and help prevent plateauing.

BARBELL BENT ROW

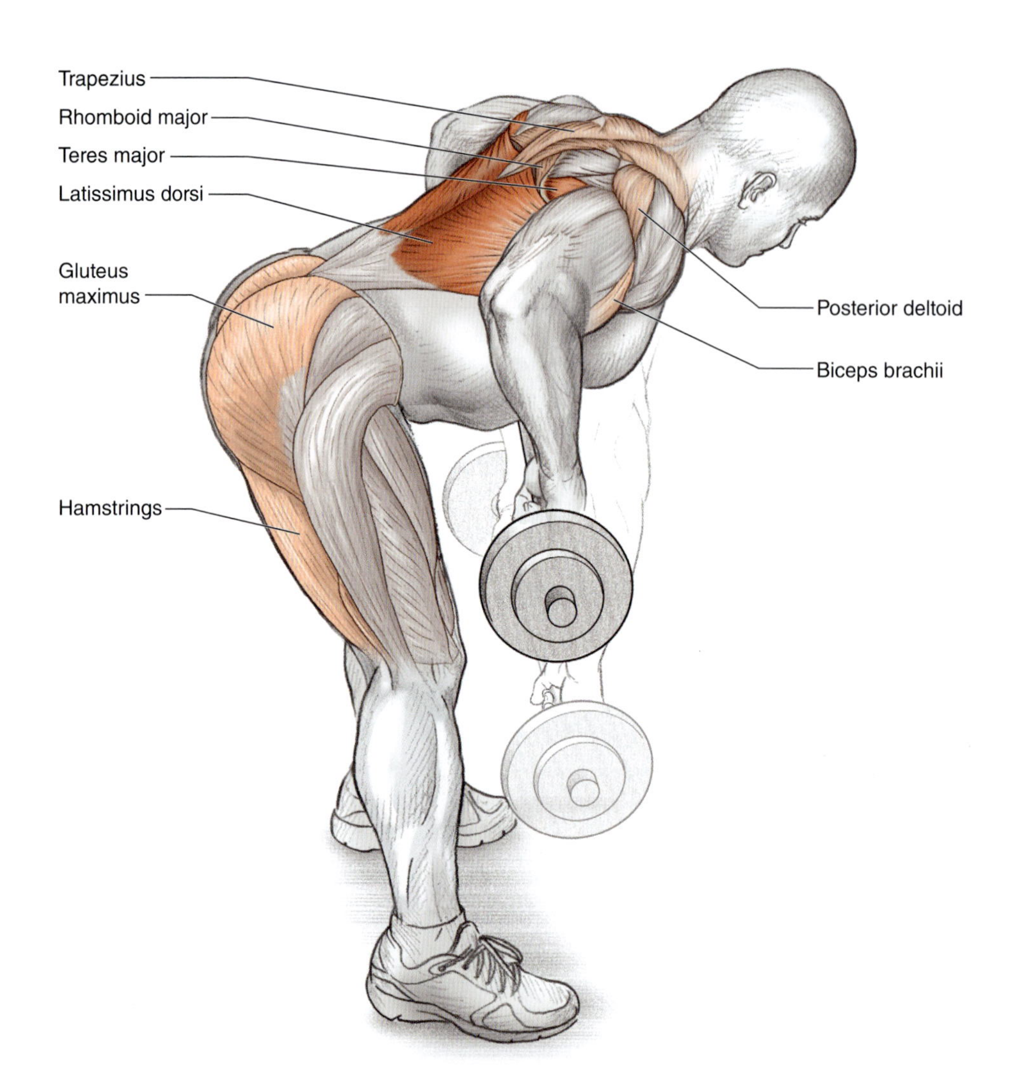

EXECUTION

1. With the barbell out in front of you, lean over at your hips while minimally bending at the knee. Keeping your back flat and chest up, lean over enough to get a strong overhand grip on the bar.
2. Pull the bar toward your lower abs. Briefly touch your abdomen without pausing, and then return the bar slowly in the eccentric phase.
3. Aim to stay in the exact same hip position so that your fully stretched arms and scapulae allow the bar to gently touch the ground at the bottom. This signifies the beginning of another rep.

MUSCLES INVOLVED

Primary: Latissimus dorsi, teres major

Secondary: Posterior deltoid, rhomboid, lower and middle trapezius, biceps brachii, brachialis, brachioradialis, erector spinae (iliocostalis, longissimus, spinalis), gluteus maximus, hamstrings (semitendinosus, semimembranosus, biceps femoris)

EXERCISE NOTES

As your scapulae protract on the way down, keep your chest up so that your back doesn't round and cause you to lose power. By keeping your chest up, you will also stretch your middle traps and rhomboids significantly under load, helping boost their growth probabilities. Keep your lower back flat so that you can stay strong and safe. Also, stay nice and low (flexed deeply at the hips) so that you can train your whole back, rather than training mostly your upper and middle traps and rhomboids with a more upright posture. It's called a barbell *bent* row for a reason! However, don't go so low as to hit the floor with your weight before your scapulae have been maximally protracted at the bottom for that very effective stretch. Stay just high up enough to let your scapulae fully stretch out as the weight gently touches the ground.

TIMING CONSIDERATIONS

Because of the antigravity support required of your erector spinae, glutes, and hamstrings, this exercise may not be ideal when you're at the top end of a muscle gain phase. Your higher body weight will cause lower back fatigue, limiting how many reps you can do. Otherwise, this is a great exercise to do for sets of 10 to 20 reps.

SAFETY

Avoid swinging around. Swinging has zero benefit, increases injury risk and fatigue, and degrades your ability to track your performance. Focus on technique first, and move your arms instead of your hips. This is not a hip extension exercise!

VARIATIONS

Experiment with different grip widths. Also you can pull the bar to your lower abdomen or to your lower chest for a stimulus more focused on the upper back and rear delt.

ASSISTED PULL-UP PARALLEL GRIP

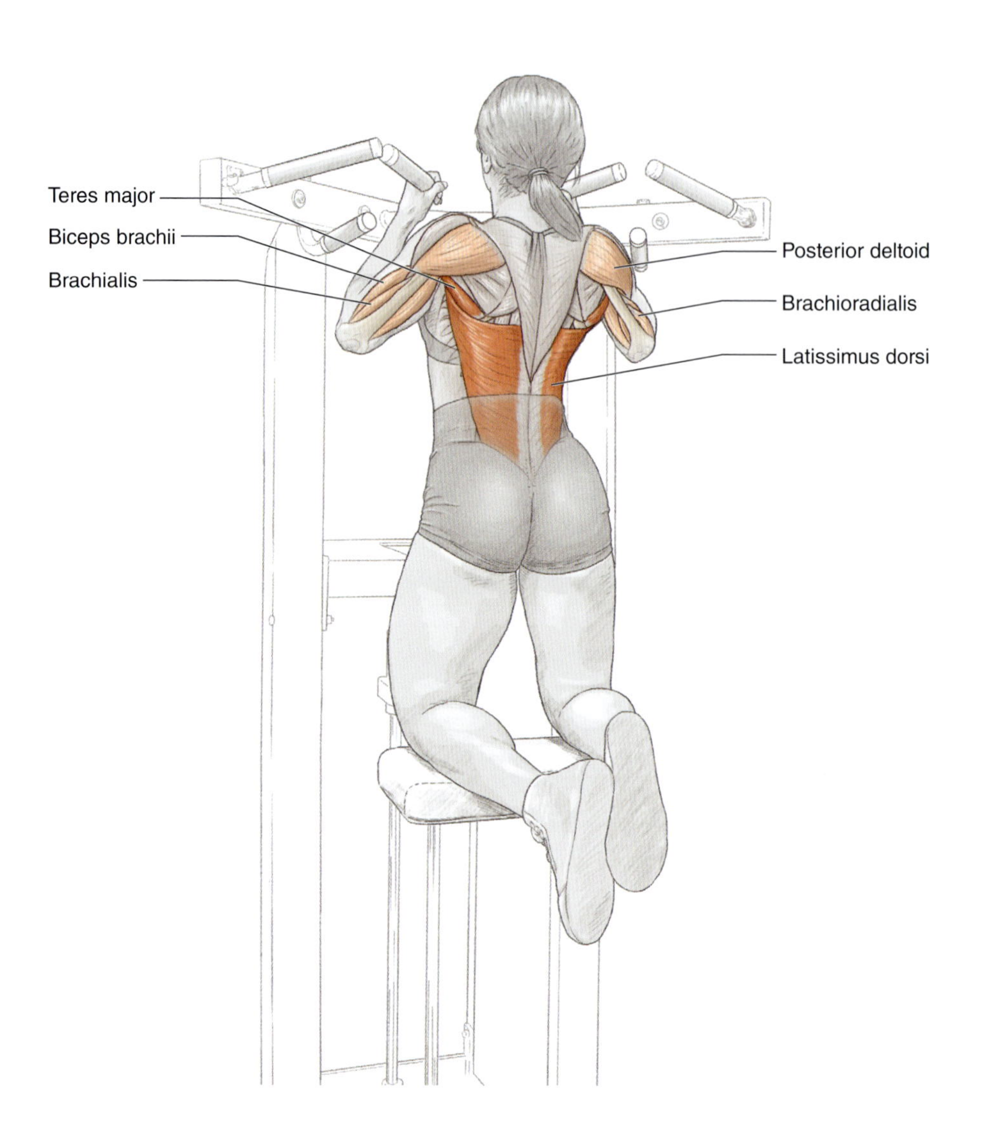

EXECUTION

1. On the assisted pull-up machine, adjust the weight so your repetitions can be smooth and controlled. Begin by gripping the pull-up bar with a parallel grip. Hang with your arms fully extended while elevating and relaxing your scapulae.
2. Gradually accelerate your ascent without jerky or sudden movements until your chin is elevated above the bar. During the pull-up, focus on bringing your elbows down by your sides to your outer hips.
3. Make a controlled and gradual descent to the starting position.

MUSCLES INVOLVED

Primary: Latissimus dorsi, teres major

Secondary: Posterior deltoid, biceps brachii, brachialis, brachioradialis

EXERCISE NOTES

Go for a deep stretch on every single rep, and keep your elbows at your sides as you pull up to engage the lats more. At the top, consider pausing for a split second to squeeze your back and reengage with the lats before slowly coming back down. This option to pause at the top is unique to the assisted pull-up and makes it a very different movement from the unassisted pull-up. This exercise is a lot more about the mind–muscle connection to the back and especially to the lats. It is less about brute force than the unassisted pull-up.

TIMING CONSIDERATIONS

This exercise is great for both weaker folks (who cannot yet do many reps of the unassisted pull-up) and for stronger and more advanced lifters (who may be seeking a deeper mind–muscle connection). For advanced lifters, performing this exercise after a few sets of unassisted or even weighted pull-ups can be very effective for stimulating hypertrophy.

SAFETY

Mind your step onto and off the machine. This is especially important when you are unracking at the end of a set, because you may be too fatigued to pull up to the right height and then step off without a lot of extra focus.

VARIATIONS

You can use any kind of grip or even hang cable grips off of the machine for variation. You can also put a yoga pad on the machine's pad to extend the top-end range of motion.

CAMBERED BAR ROW

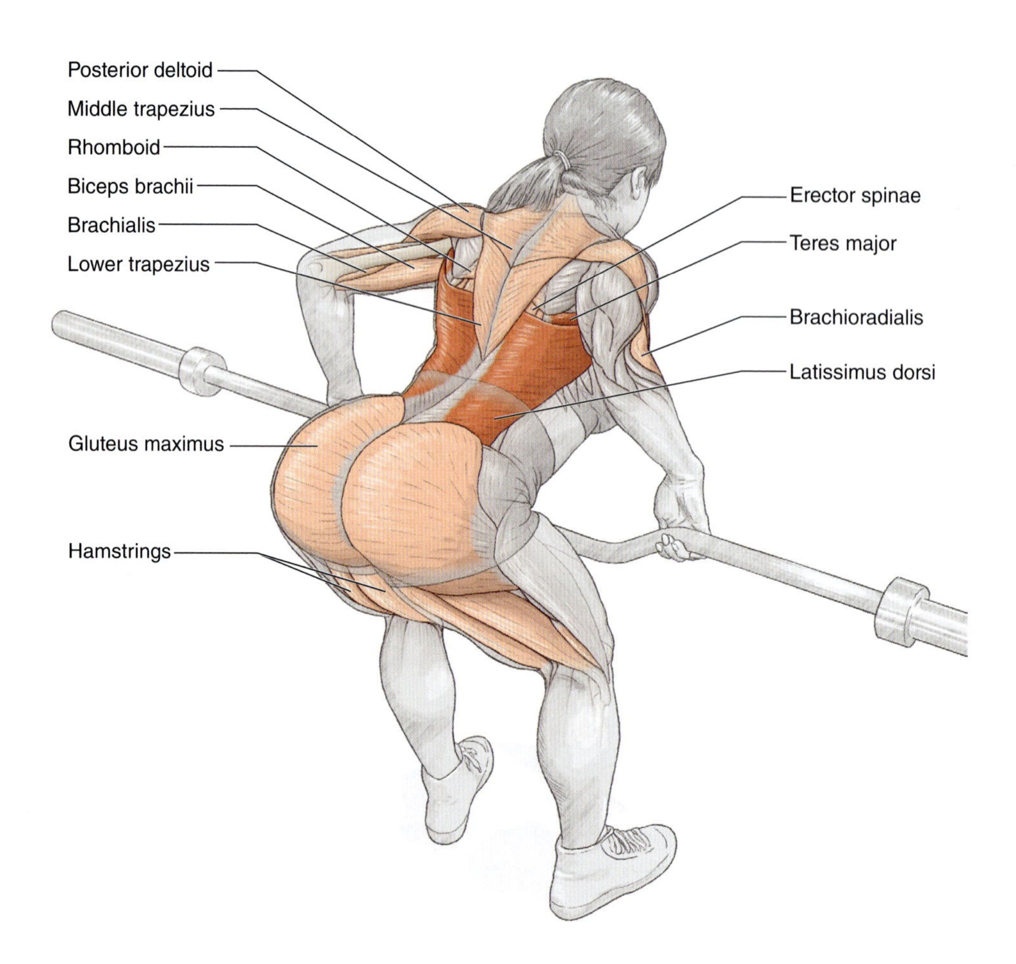

EXECUTION

1. Start with the bar in front of your body. Hinge at the hips while maintaining a minimal bend at the knee. Keep a strong overhand grip on the bar. Then lean over, maintaining a flat back and elevated chest.
2. Pull the bar toward the lower abdomen and briefly make contact with the abdomen. Then ensure a controlled eccentric phase when returning the bar to its initial position. It is important to maintain the same hip position throughout the exercise to maximize the stretch of the fully extended arms and scapulae.
3. At the bottom of each repetition, gently touch the bar to the ground before initiating the next repetition.

MUSCLES INVOLVED

Primary: Latissimus dorsi, teres major

Secondary: Posterior deltoid, rhomboid, lower and middle trapezius, biceps brachii, brachialis, brachioradialis, erector spinae (iliocostalis, longissimus, spinalis), gluteus maximus, hamstrings (semitendinosus, semimembranosus, biceps femoris)

EXERCISE NOTES

Lean to the degree that you are able to touch the bar to your lower abdomen each rep. This portion of the lift is a big deal in this exercise, so you shouldn't miss out on it. As you pull, try to pull back toward the lowest part of your abdomen, as this often results in the best peak contraction for the muscles of the back. As soon as you reverse direction, control the eccentric phase from the top, even when your elbows are still behind your back.

TIMING CONSIDERATIONS

That pinch at the top end crushes your mid-back (rhomboids, erector spinae, middle of the traps, and teres major). Because it also passes the leverage point of your lats, choose this as an overall back exercise, not just as a lat-specific back exercise.

SAFETY

It can become tempting to use momentum and hip movement at the top to complete the rep, but resist. Use only your target muscles to do the work. This will create more growth with less fatigue.

VARIATIONS

If you don't have access to a cambered bar, dumbbells and machines can replicate this movement to some degree.

UNDERHAND PULL-DOWN

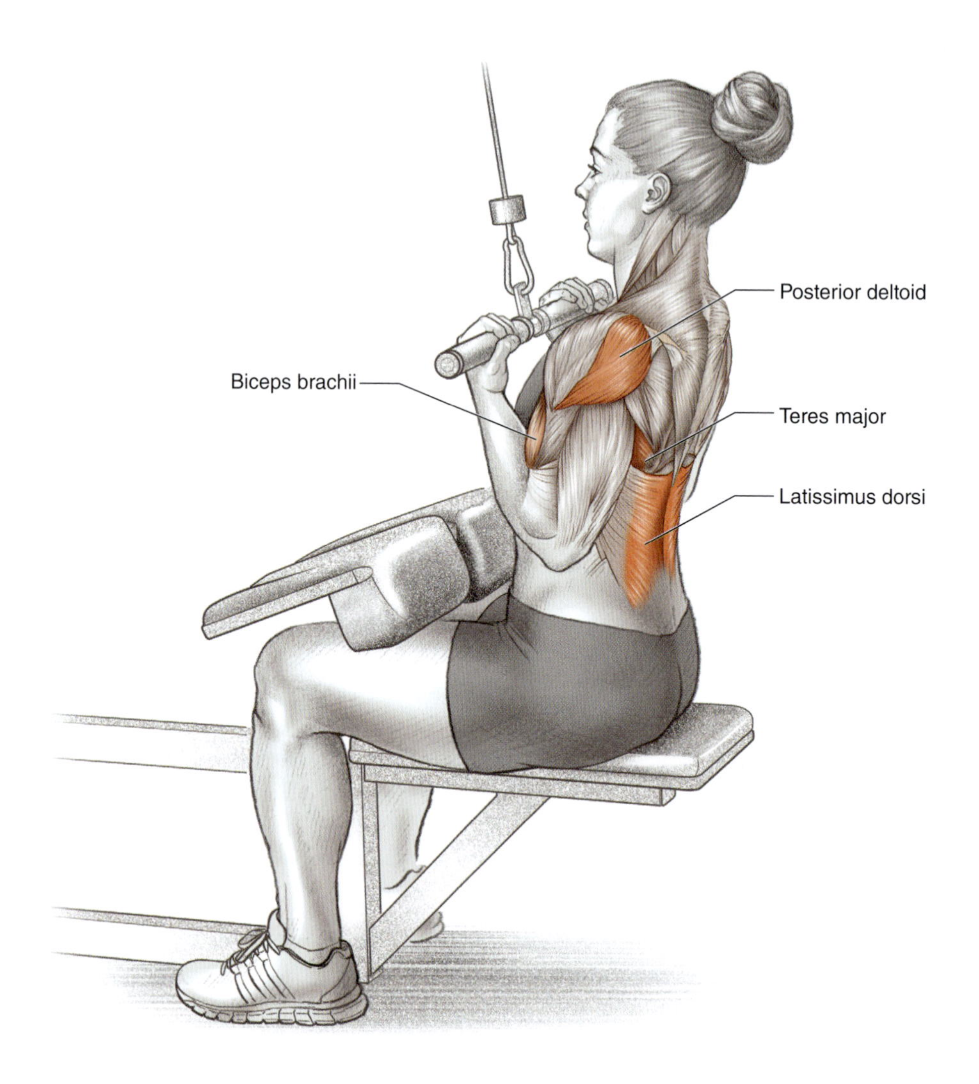

EXECUTION

1. Sit down at the cable pull-down machine with the bar attached to a high pulley. To begin the exercise, use an underhand grip on the bar with your arms fully extended. Elevate and relax your scapulae.
2. Gradually accelerate the descent (without jerky or sudden movements) until the bar touches your clavicles. As you pull down, focus on bringing your elbows toward your sides instead of flaring them out.
3. Ensure a controlled and gradual ascent back to the starting position.

MUSCLES INVOLVED

Primary: Latissimus dorsi, teres major, posterior deltoid
Secondary: Biceps brachii, brachialis, brachioradialis

EXERCISE NOTES

Keep your elbows in and flush to your sides on the way down for the best lat stimulus. When coming down, touch the bar where it is most comfortable and where your mind–muscle connection with the lats is the best. If this means you touch the clavicles while sitting straight up or the lower chest while arched back a bit, that's fine. There's no wrong answer here. Even though it's not very comfortable, don't skip the mega stretch at the top of this exercise.

TIMING CONSIDERATIONS

Anyone can benefit from this exercise, but it's a great movement for focusing on the mind–muscle connection. More advanced lifters might get more out of it by intentionally squeezing the lats down at the bottom and riding the eccentric phase with high lat activity on the way up.

SAFETY

If the straight bar is hard on your wrists or elbows, see if an EZ bar attachment feels better.

VARIATIONS

In addition to an EZ bar handle, there are many special attachments you can use for variation, such as the Prime handles or MAG Grips.

CABLE FLEXION ROW

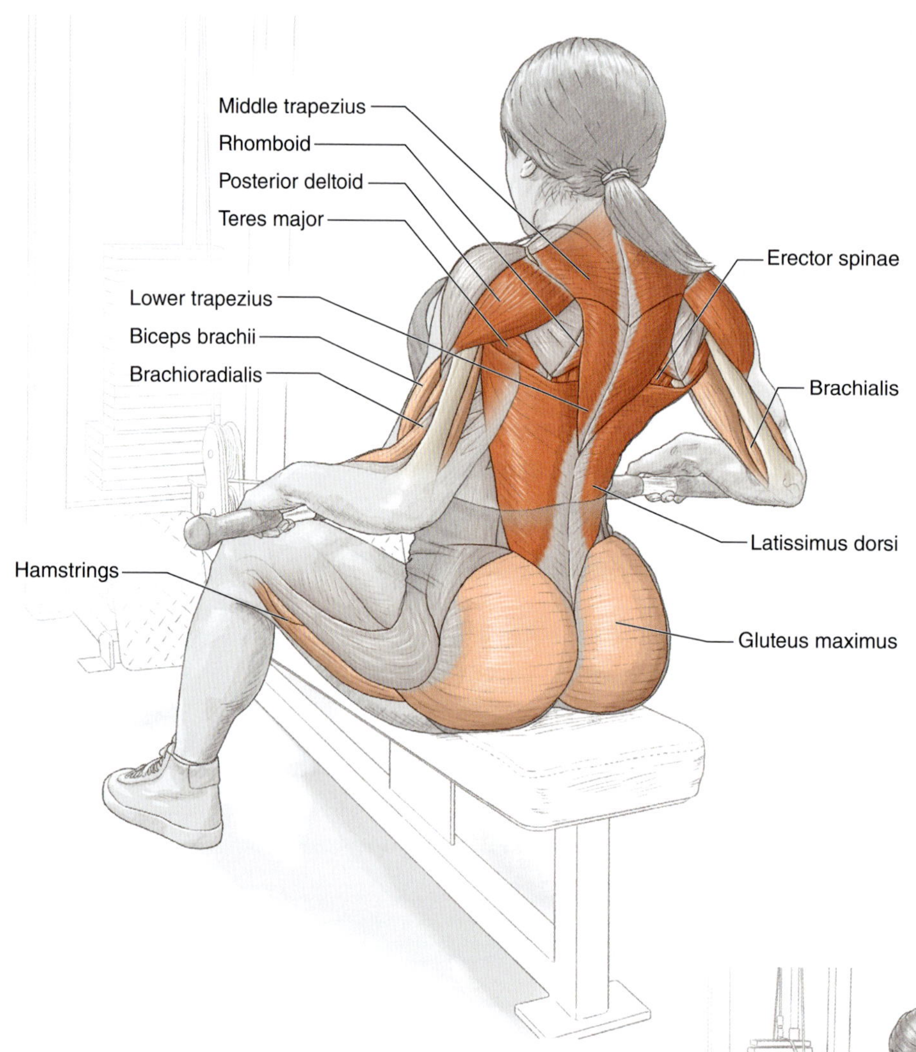

EXECUTION

1. Begin by sitting a few feet away from a cable pulley with your torso vertical and your back flat. Grab a lat pull-down bar attached to a low pulley, and take an overhand grip.
2. Round your back and protract your scapulae as you slowly stretch out. Stop the stretch when you can no longer descend any further.

Starting position.

3. Gently begin the ascent by arching your back, flexing at your elbows, and retracting your scapulae.
4. Finish the movement with a vertical torso and your chest up, your scapulae back as far as they can go, and the bar touching your abdomen.

MUSCLES INVOLVED

Primary: Latissimus dorsi, erector spinae (iliocostalis, longissimus, spinalis), teres major, posterior deltoid, rhomboid, lower and middle trapezius

Secondary: Biceps brachii, brachialis, brachioradialis, gluteus maximus, hamstrings (semitendinosus, semimembranosus, biceps femoris)

EXERCISE NOTES

You may want to hold your breath on the way down and only breathe out at the halfway point of the concentric phase. This will help you maintain some core tightness. However, don't try to keep the back flat; actively get into as much spinal flexion as you can. If you go for a huge stretch at the bottom, almost every part of your back will grow, even though you may only be able to use lighter weight to do this. When arching up at the top, make sure your lower back is still at most perpendicular to the ground. There is no benefit to leaning back, except that it makes the movement needlessly easier.

TIMING CONSIDERATIONS

This exercise, when done properly, will cook your spinal erectors in the best kind of way and stimulate loads of growth and fatigue. To that end, be careful. Plan when and how much you'll need your spinal erectors for other back and leg training in your current program so that they can recover on time and not limit you later in your training week.

SAFETY

Start out light (20 or 30 reps at the most), and keep the movement smooth on both the eccentric and concentric phases. This will lower the risk of injury to your lower back. In fact, because this exercise builds lower back muscle and dynamic lower back movement competency, it likely reduces overall risk of injury or the risk of developing nagging lower back pain from the rest of your activities.

VARIATIONS

You can use any kind of handles for variation. You can also do this exercise with dumbbells, barbells, and even on the Smith machine. If you put the chest pad on your abdomen instead of your chest, you can perform this exercise on a row machine as well.

LAT PRAYER

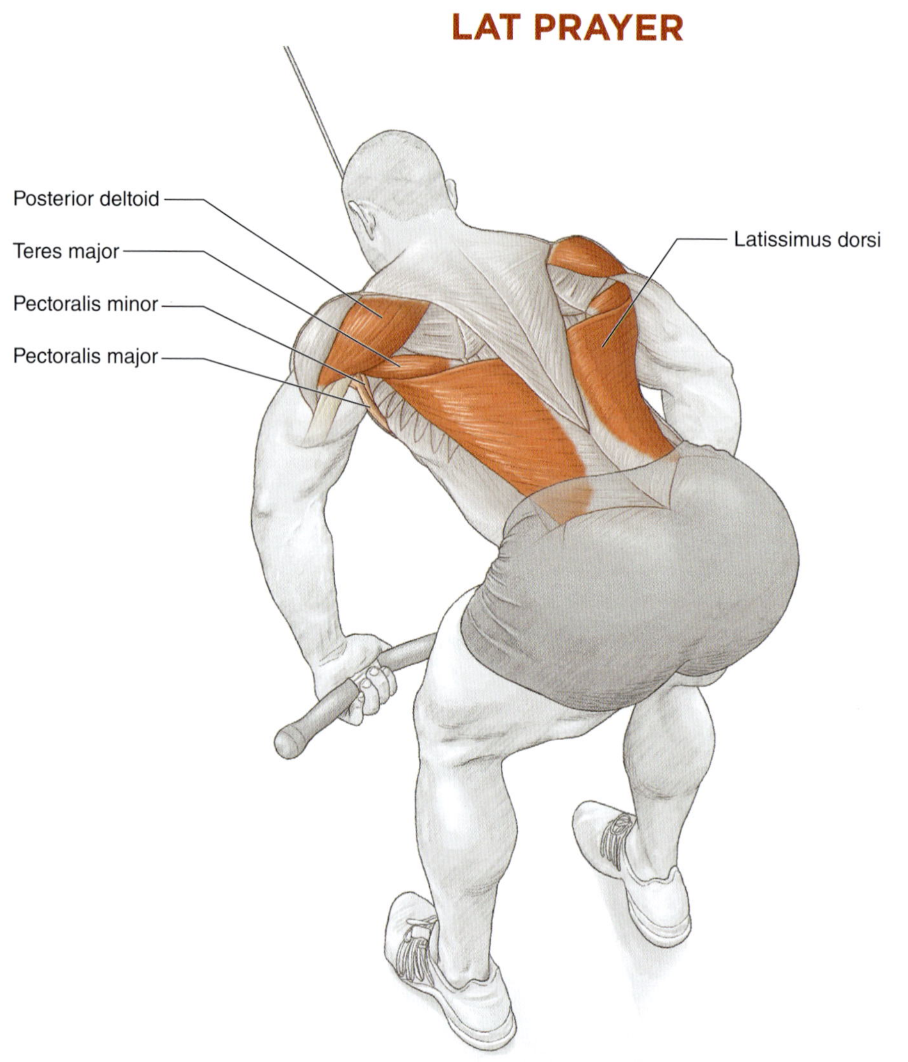

EXECUTION

1. Set the cable height of the cable machine at around the same height as your clavicles when you're standing. Grip whatever cable attachment you're using.
2. Back up a bit, and lean your upper body forward at the hips so that your arms are fully outstretched at the top. The cable handle should be a bit above your shoulders, and your lats should be fully stretched.

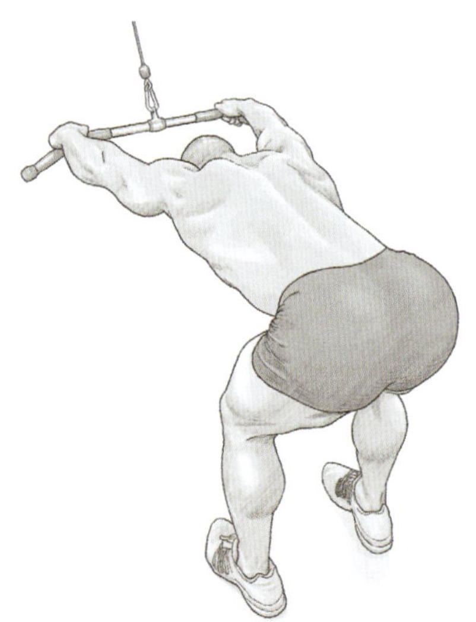

Starting position.

3. Pull the bar down in an arcing motion toward your feet. As the cable nears your face, lift your chest a bit to make sure the cable doesn't hit you.
4. Gently touch your thighs or go just outside of them with the cable attachment. Then slowly return to the starting position (the full stretch).

MUSCLES INVOLVED

Primary: Latissimus dorsi, teres major, posterior deltoid

Secondary: Pectoralis major, pectoralis minor

EXERCISE NOTES

Because each rep takes so long to perform, you can really benefit from controlling the eccentric phase. As you enter the last third of the eccentric phase, consider leaning your chest down a bit. This will put more of a loaded stretch on the back muscles and rear delts at the top before you reverse directions. At the bottom third, when you lift your chest up, only lift your chest enough to just barely clear the cable from hitting your face. Lifting the chest more than this will degrade the force curve of this exercise. It will also limit you more at the peak contraction than is ideal for growth by robbing you of more stimulus at the loaded stretch.

TIMING CONSIDERATIONS

You can perform this exercise first in a session, but it also works great after a compound vertical pull move.

SAFETY

Rapid reversal is a bad idea, because there is a lot of load to transduce through muscles when they are at their most outstretched positions and when the glenohumeral joint is at its end range. Always be gentle in the reversal from eccentric to concentric on this exercise. If you can't obtain a full stretch because of shoulder discomfort, try a different grip width or cable attachment. If that doesn't work, stop at the highest range of motion where your shoulders still feel comfortable.

VARIATIONS

A neutral close grip is the best grip to hit the lats in most cases, but you can always try another grip that's comfortable for you. There are really no wrong answers with this exercise—just slight emphasis shifts between the many muscles involved.

MACHINE ROW

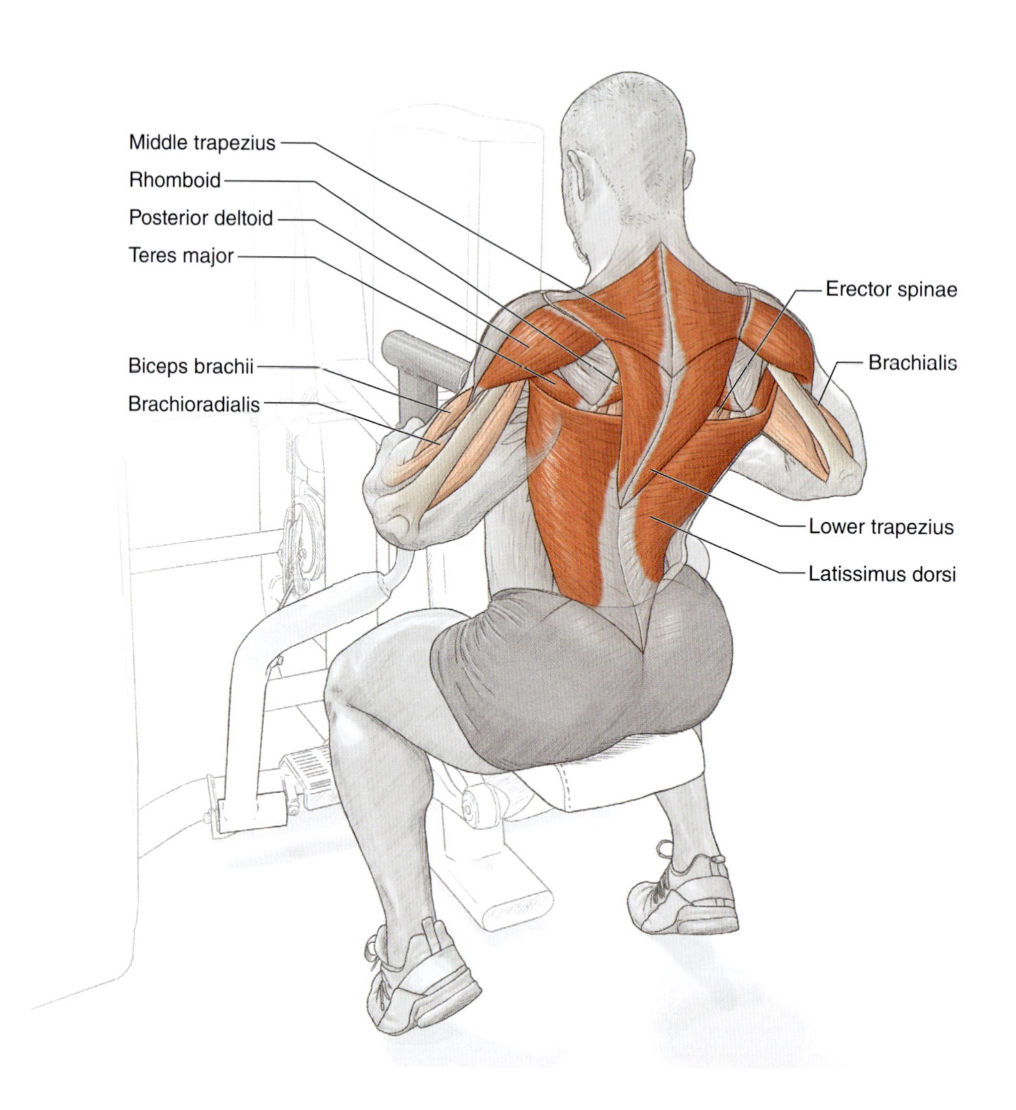

EXECUTION

1. Begin by sitting onto the pad of the machine with your torso vertical, your back flat, and your abdomen on the pad.
2. Round your back and protract your scapulae as you slowly stretch out. Stop the stretch when you can no longer descend any further.
3. Gently begin the ascent by arching your back, flexing at your elbows, and retracting your scapulae.
4. Finish the movement with your torso vertical and your chest up, your scapulae back as far as they go, and your elbows as far back as they can go while your abdomen is still on the pad.

MUSCLES INVOLVED

Primary: Latissimus dorsi, teres major, posterior deltoid, rhomboid, lower and middle trapezius

Secondary: Biceps brachii, brachialis, brachioradialis, erector spinae (iliocostalis, longissimus, spinalis)

EXERCISE NOTES

It's rare to be able to emphasize the peak contraction with free weights because they bias the loaded stretch so much in their force curves. On a machine, the force curve is often more even, or makes the peak contraction easier. It might be worth your time to pause at the top of the motion with your elbows, scapulae, and whole back crunched for a second or two before descending. When you move back down, ride out the eccentric phase slowly, and let it pull you into scapular protraction and spinal flexion at the bottom. This will reduce how much weight you're lifting, but nobody cares about how much you can machine row!

TIMING CONSIDERATIONS

A rowing machine—especially one with chest support (most of them)—is great when you're fatigued. It is also useful when your lower back is fatigued and needs to be fresh to support another muscle group's training soon. You can just sit down, plug in, and grow your back without having to spend much physical or mental energy on supporting your body against gravity.

SAFETY

Machine rows are very safe, but pick your grips and chair heights to get the least shoulder joint discomfort you can. How high or low the chair height is and which grips you pick can make a huge difference in how this exercise feels.

VARIATIONS

You can do traditional chest-supported rows, where you lay the pad across your chest and you don't round your mid and lower back at the bottom. You can also try raising the seat height and placing your abdomen on the pad instead. This will clear your chest and allow it to get you into spinal flexion at the bottom of the row. Both are great alternatives. Consider using one for a few months and then trying the other.

ONE-ARM CABLE HIGH-ROW

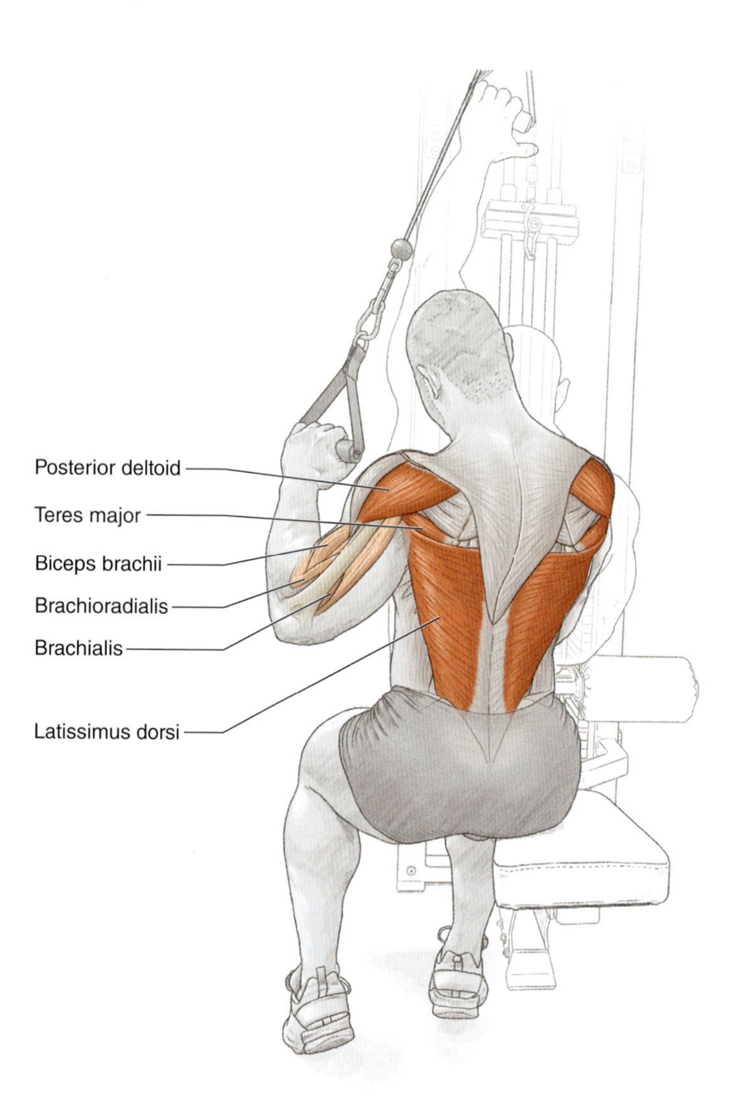

EXECUTION

1. At the lat pull-down station, grab the handle with one arm and sit off-center by about 6 inches (15 cm). This will create more pulling distance for your working arm. Make sure the movement begins at a maximum stretch for your lats, rear delts, biceps, and scapulae. Your trunk should also tilt somewhat toward the handle.
2. As you pull down, bring your elbow as close to your hip as possible, slightly tilting back in the other direction and depressing the scapulae.
3. Slowly return to the top stretch, and repeat. Switch sides only after completion of the set.

MUSCLES INVOLVED

Primary: Latissimus dorsi, teres major, posterior deltoid
Secondary: Biceps brachii, brachialis, brachioradialis

EXERCISE NOTES

You can do this exercise while sitting in the middle of the pull-down station. However, if you sit off-center, you can pre-stretch the lats even more and potentially present them with an even bigger hypertrophy stimulus. At the top end of this motion, you'll feel like it's more of a pull-down as you'll be very vertically oriented. At the bottom, you might want to lean backward and off to the side a bit, replicating a rowing motion to squeeze the lats at their lowest fibers. For shoulder comfort and more range of motion for the muscles, consider pronating at the top of the stretch and then, as you pull your elbow down and back toward your hip, supinating. At the top, tilt into the cable attachment; at the bottom, tilt away for even more lat stretch and engagement.

TIMING CONSIDERATIONS

This exercise is probably best suited for more advanced lifters who will benefit more from the mind–muscle connection this exercise requires. For many beginner and intermediate lifters, it is probably a wiser investment to spend time on variations of pull-ups and rows.

SAFETY

Momentum on this movement is not wise. It is not worth injury and irritation to the lat, delt, triceps, and glenohumeral joint.

VARIATIONS

You may have seen this exercise demonstrated with a restricted top-end range of motion, which does not allow the scapula to fully elevate at the top. This style limits the back muscles and especially the lats from stretching under load as much as they could. Thus it is probably not great for promoting growth.

CHAPTER 4

Arms

In combination with the chest, shoulders, and back, the arms are the final group of muscles to consider when developing proper upper body aesthetics for bodybuilding. For bodybuilders, well-developed and symmetrical arm muscles are not only aesthetically alluring, but development of the arm muscles is also crucial for optimizing performance.

The elbow joint—formed by the articulation between the humerus (upper arm) and the radius and ulna (forearm)—is a hinge joint that provides the fulcrum for most arm exercises. The elbow joint is further divided into two distinct articulations: the humeroulnar joint and the humeroradial joint. The humeroulnar joint is formed by the trochlea of the humerus and the trochlear notch of the ulna. Its primary function is to allow for flexion and extension of the forearm at the elbow. Typically it is the articulation between the humerus and the ulna that limits the ability to fully extend the elbow; it can even cause hyperextension at the elbow. The humeroradial joint is formed by the capitulum of the humerus and the head of the radius. It allows for flexion and extension. It also contributes to pronation and supination of the forearm when combined with the proximal radioulnar joint.

The muscles of the arm can be broadly categorized into two main regions: the anterior (front) compartment (which contains the flexor muscles), and the posterior (back) compartment (which houses the extensor muscles). The anterior compartment of the upper arm primarily comprises the biceps brachii, brachialis, and brachioradialis, which collectively contribute to flexion and supination of the forearm. The posterior compartment of the upper arm is dominated by the triceps brachii, responsible for forearm extension and elbow joint stabilization. The forearm is also divided into similar components.

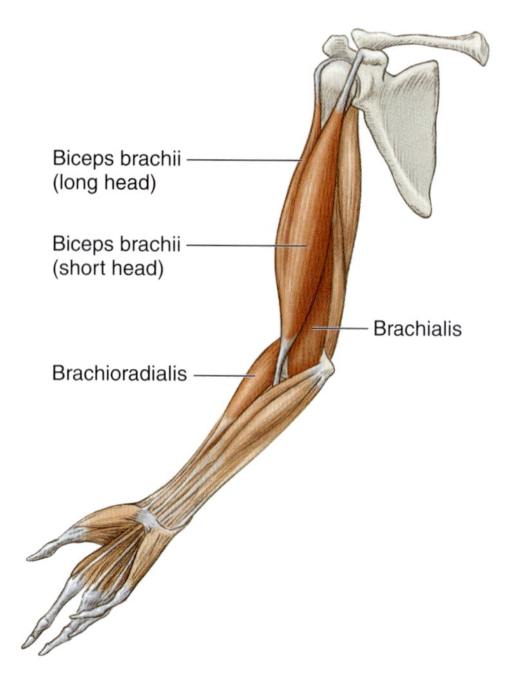

Figure 4.1 Biceps, brachialis, and brachioradialis.

The biceps brachii (figure 4.1), also known as the "biceps," is located in the anterior compartment of the upper arm. It is a two-headed muscle, with each head originating from a different point on the scapula and converging to a single tendon that inserts onto the radial tuberosity of the radius bone in the forearm. The long head originates from the supraglenoid tubercle of the scapula above the shoulder joint. The short head originates from the coracoid process of the scapula. The main function of the biceps is to flex the elbow and to rotate the forearm into a supinated position. The biceps curl and its multiple variations will target this muscle considerably. Given that the long head of the biceps originates above the shoulder joint, the biceps can contribute to shoulder flexion to a small degree.

In addition to the biceps brachii, the other two muscles in the anterior compartment of the upper arm include the brachialis and brachioradialis. The brachialis muscle is a flat, rectangular muscle located beneath the biceps brachii. It originates from the distal half of the humerus and inserts onto the coronoid process and the tuberosity of the ulna in the forearm. Its main function is to create flexion at the elbow. The brachioradialis originates from the supracondylar ridge of the humerus and inserts onto the styloid process of the radius. Its main function is to flex at the elbow, but it also contributes to forearm pronation and supination, particularly when the elbow is in a flexed position.

The triceps brachii (figure 4.2), also known as the "triceps," is a large, three-headed muscle located in the posterior compartment of the upper arm. The long head originates from the scapula just below the shoulder joint. The lateral head originates from the posterior surface of the humerus. The medial head originates from the posterior surface of the humerus. All three heads of the triceps converge into a single tendon that inserts into the olecranon process of the ulna in the forearm. The primary function of the triceps is to extend the elbow. This is critical for pushing and pressing movements.

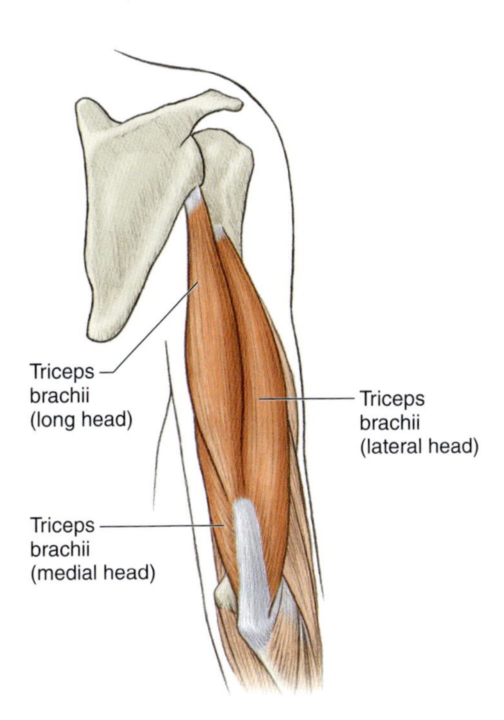

Figure 4.2 Triceps.

Finally, the forearm (figure 4.3) contains about 20 individual muscles, with a similar arrangement to that of the upper arm. The flexor muscles are in the anterior compartment facing the palm side, and the extensor muscles are in the posterior compartment facing the reverse side. The anterior compartment contains seven to eight muscles in total, with two to three muscles responsible for wrist flexion (palmaris longus—not everyone has a palmaris longus, flexor carpi radialis, and flexor carpi ulnaris); three muscles responsible for finger flexion (flexor digitorum superficialis, flexor digitorum profundus, and flexor pollicis longus); and two muscles responsible for forearm pronation (pronator teres and pronator quadratus).

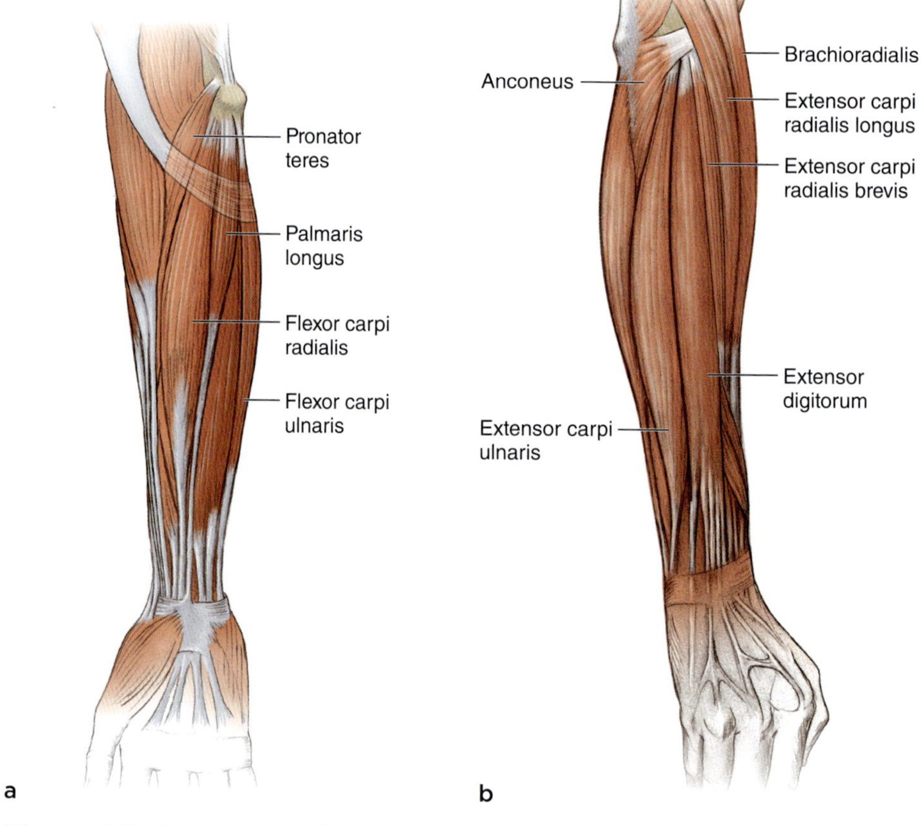

Figure 4.3 Forearm muscles: (*a*) flexors; (*b*) extensors.

The posterior compartment contains 12 muscles, with three muscles responsible for wrist extension (extensor carpi radialis longus, extensor carpi radialis brevis, and extensor carpi ulnaris); six muscles responsible for finger extension (extensor pollicis longus, extensor pollicis brevis, abductor pollicis longus, extensor digitorum, extensor indicis, and extensor digiti minimi); and one muscle responsible for forearm supination (supinator). The remaining two muscles in the posterior compartment include the anconeus and brachioradialis. The anconeus muscle is a very small muscle that mainly functions in assisting the triceps with elbow extension. The brachioradialis is a unique exception in that the muscle is located in the posterior compartment but is mainly responsible for forearm flexion.

BARBELL CURL

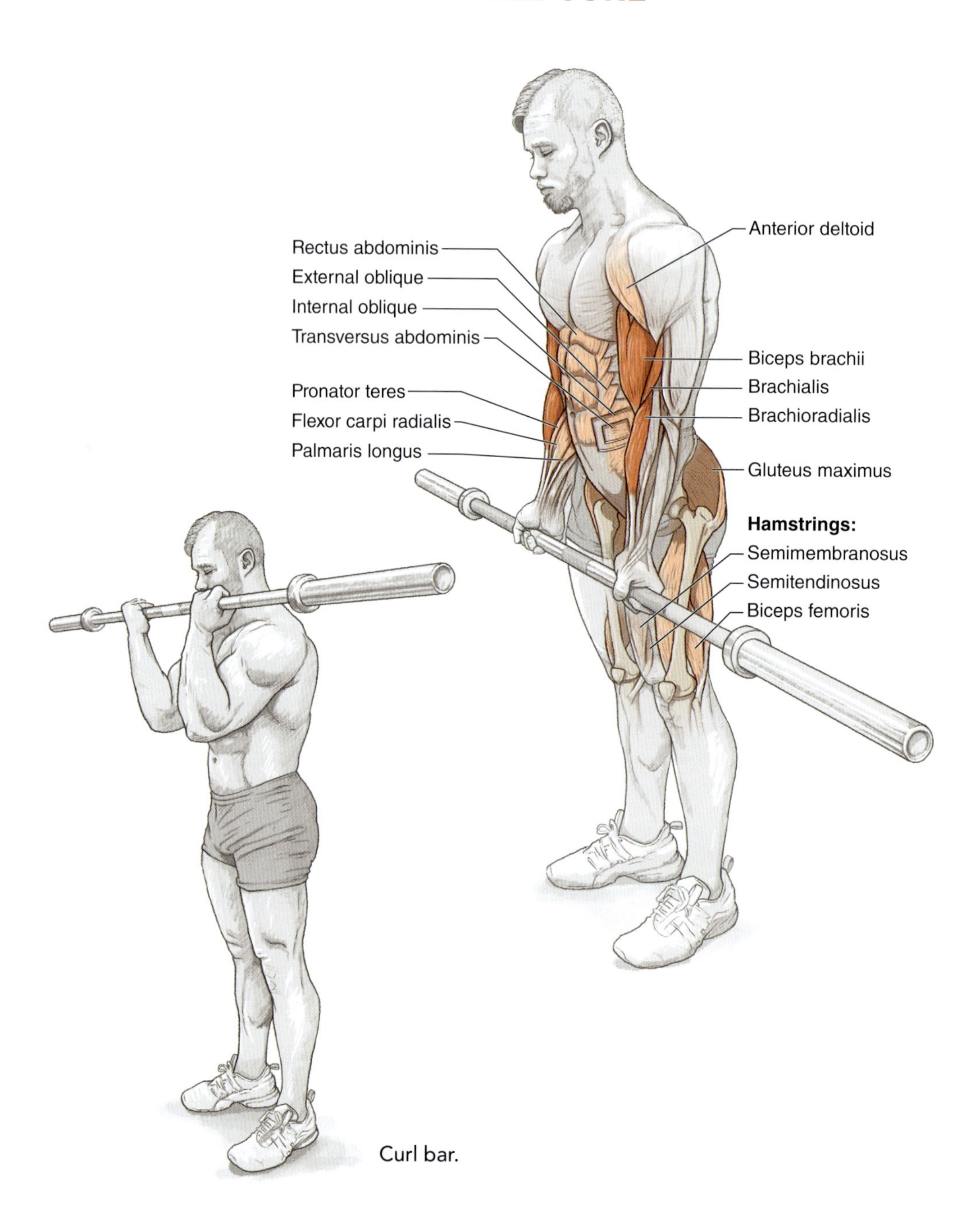

Curl bar.

EXECUTION

1. Stand up straight with your feet in a comfortable position, approximately shoulder-width apart. Begin with your hands underneath the bar at just inside shoulder width, with your arms straight down, and your chest up.

2. Curl the bar up in an arcing motion until your forearms touch your biceps. As you curl up, keep the elbows from flaring out by actively pushing them inward.
3. Slowly return to the starting position.

MUSCLES INVOLVED

Primary: Biceps brachii, brachialis, brachioradialis

Secondary: Pronator teres, flexor carpi radialis, palmaris longus, anterior deltoid, rectus abdominis, transversus abdominis, internal oblique, external oblique, erector spinae (iliocostalis, longissimus, spinalis), gluteus maximus, hamstrings (semitendinosus, semimembranosus, biceps femoris)

EXERCISE NOTES

Don't cut out the bottom part by reversing the motion 10 or 15 degrees before elbow extension. This can rob you of an extra stretch under load stimulus. Because you will experience the highest forces when your forearms are parallel to the floor in the middle of the motion, this is where you should focus on the most eccentric slowing and control of the bar (which is exactly where you'll feel like rushing when you get tired). To enhance the stretch at the bottom, lift your chest, and lean back at the bottom of the exercise. This will pre-stretch your biceps and give them a better stimulus.

TIMING CONSIDERATIONS

This foundational exercise is great for flexible people, beginners, and intermediates. If you can master the barbell curl, all other curls will be more effective, so it's a worthy investment.

SAFETY

Feel your grip and elbow position, as this exercise might irritate your wrists, elbows, or shoulders, depending on which one has to be in the least ideal position to accommodate the others. If your hand position doesn't work, try using an EZ bar. More muscular and advanced lifters often have to move to an EZ bar. Additionally, this exercise can impose significant stress on the lower back. That isn't a bad thing, but it might interfere with lower back recovery for other exercises in your program, such as squats and deadlifts. To reduce lower back stress, wear a lifting belt while doing barbell curls (even if it looks a bit curious to some observers).

VARIATIONS

Raise your upper arms during this movement or keep them completely perpendicular to the ground the whole time. Both variations can be effective. You can even do one for a few months, and then replace it with the other.

EZ BAR CURL

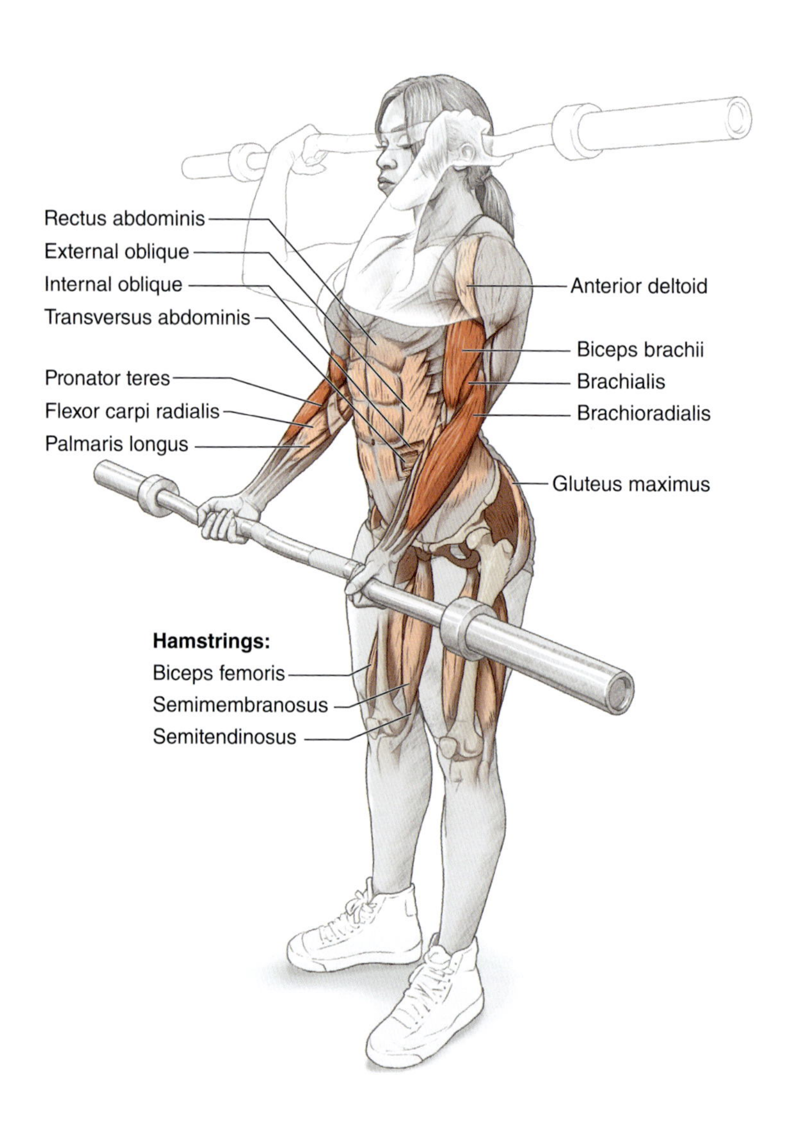

EXECUTION

1. Stand straight with your feet in a comfortable position, approximately shoulder-width apart. Begin with your hands on either the outer or inner bends of the EZ bar, with your arms straight down and your chest up.
2. Curl the bar up in an arcing motion until your forearms touch your biceps. As you curl up, keep the elbows from flaring out by actively pushing them inward.
3. Slowly return to the starting position.

MUSCLES INVOLVED

Primary: Biceps brachii, brachialis, brachioradialis

Secondary: Pronator teres, flexor carpi radialis, palmaris longus, anterior deltoid, rectus abdominis, transversus abdominis, internal oblique, external oblique, erector spinae (iliocostalis, longissimus, spinalis), gluteus maximus, hamstrings (semitendinosus, semimembranosus, biceps femoris)

EXERCISE NOTES

If you try to actively push your elbows in toward your chest during this movement, you can enhance the mind–muscle connection of the biceps (especially if you actively slow down during the middle portion of the exercise's eccentric phase). If you can find a bench where you can set the bar between sets, you can avoid having to deadlift it every time before and after a set; this can save energy for the curl itself.

TIMING CONSIDERATIONS

This exercise is a great variation to straight bar curls of any kind; straight bar curls might irritate your wrists, elbows, or shoulder joints.

SAFETY

This exercise is very safe, especially for sets of 10 or more reps. Do not, however, start swinging on the concentric phase to try to use more weight than your forearm flexors can lift on their own.

VARIATIONS

Vary the inner and outer EZ grip hand positions. There's no wrong answer here, so feel free to use both in a sequenced manner, switching every few months.

INCLINE DUMBBELL CURL

EXECUTION

1. Set an incline bench to roughly a 45- or 60-degree angle. Lie supine on the incline bench.
2. Pick up a dumbbell in each hand, and hold your legs close together in front of you, with your feet flat on the floor. If your feet do not reach the floor, you may keep your toes planted on the floor, or you may consider putting a weight plate underneath your feet.
3. Let the dumbbells come all the way down and back, while keeping your back slightly arched and your chest up to the ceiling. Then curl the dumbbells up until your forearms touch your biceps.
4. Slowly return the dumbbells down for a maximum stretch while your chest stays up the whole time.

MUSCLES INVOLVED

Primary: Biceps brachii, brachialis, brachioradialis

Secondary: Pronator teres, flexor carpi radialis, palmaris longus, anterior deltoid

EXERCISE NOTES

Set the seat angle for a big stretch and shoulder comfort. If you set the seat too vertically, you won't get a big stretch in the biceps; if you set it too horizontally, you might irritate your shoulder joints. Start with a 60-degree seat angle and lower it as much as you can while still feeling comfortable in your shoulder joints. To support as much loaded stretching as possible, don't just restart the concentric as soon as your arms are near the bottom position. Instead, actively reach back behind you on each rep's eccentric end, allowing maximum stretch before the concentric of the next rep begins. When curling, you should have your knees together; if you spread them out, the dumbbells will hit your legs on the way up. Lastly, make sure to slow and control the eccentric phase a lot on this exercise. This control will not tire out your lower back (as it might in standing curls), so this is the ideal place to focus on a quality eccentric phase.

TIMING CONSIDERATIONS

This is a very stable exercise with a huge loaded stretch exposure and a ton of range of motion. It's about as effective as biceps exercises get. If you want the best long-term growth for your biceps, use it at the end of a tough training phase or during a biceps specialization block when you need the highest stimulus possible. For an extra bit of stimulus, try some myo-reps with this exercise: Do as many reps as you can, rest the dumbbells on your knees for five to 10 seconds, and then do as many reps as you can again; repeat this two or three times in a row. Ouch!

SAFETY

Rush the eccentric phase at your own risk. If you use too much weight and get too dynamic, you could risk injury in the loaded stretch position of this exercise.

VARIATIONS

For variety, try to hold the dumbbells on the insides of the handles versus the outside. When holding on the inside, you'll be able to use more load and focus more on the dumbbell's path and eccentric control. When holding on the outside, you'll have to focus on lifting your pinkie fingers up to train some of the supination component of the biceps function. This will limit loads and other technical focus opportunities, but it will also provide a unique biceps stimulus.

FREEMOTION STRETCH CURL

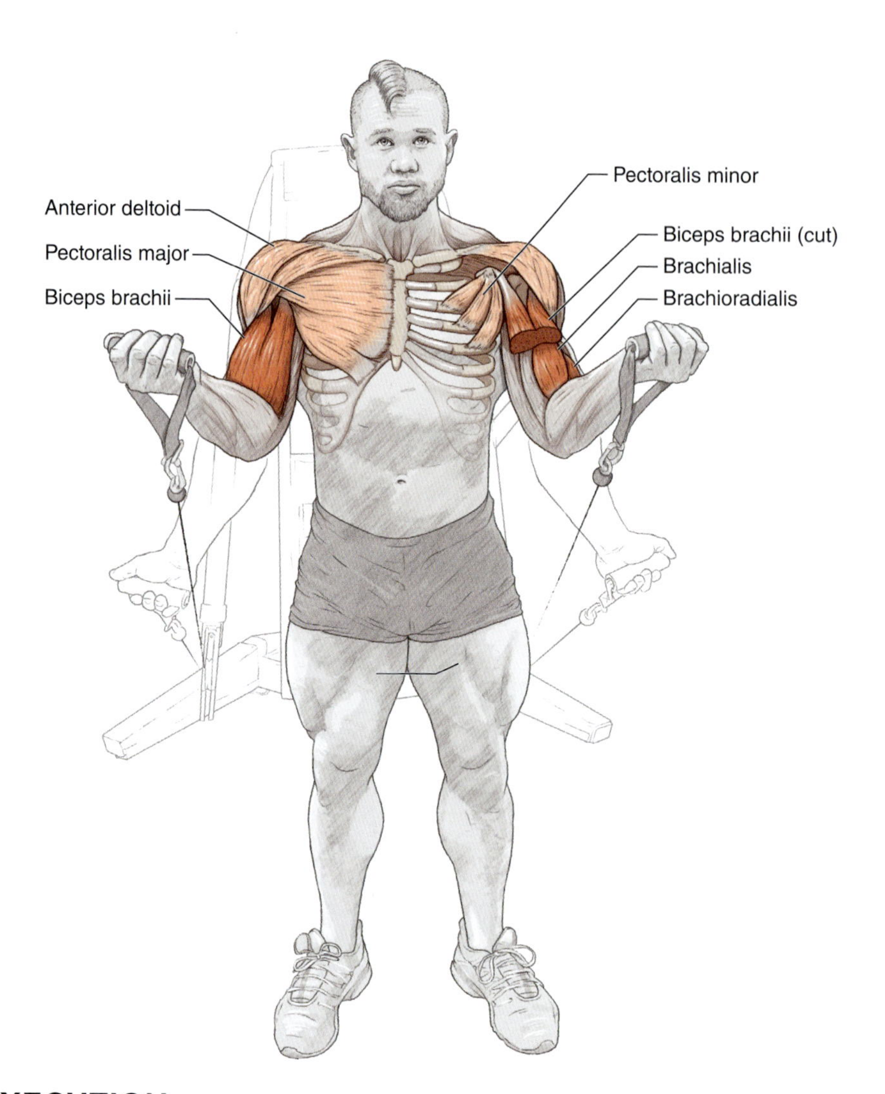

EXECUTION

1. Set the cable positions to a setting somewhere between your feet and your hips, and set the width anywhere between shoulder width and as far out as you'd like to try. You can stand using a staggered stance (with one foot slightly in front of the other and both feet flat on the ground) or with your feet shoulder-width apart—whichever is more comfortable.

2. Facing away from the machine, grab both handles and walk forward so that you can pull the slack out and feel a lot of tension at the very stretched position.
3. With your chest up, curl the handles until your forearms touch your biceps. Then return to the starting point.

MUSCLES INVOLVED

Primary: Biceps brachii, brachialis, brachioradialis

Secondary: Pronator teres, flexor carpi radialis, palmaris longus, anterior deltoid, pectoralis major, pectoralis minor

EXERCISE NOTES

Using a staggered stance can potentially increase stability while curling. Keep a big chest at the bottom, and let your hands move way back; this will put the biggest stretch possible on your biceps while not going so far as to irritate your shoulder joints. If you're not getting a lot of tension at the bottom stretch, walk forward a few steps with the handles in your hands. The farther back your hands are (with a high chest) at the bottom, the better; however, you don't need to move your upper arms up at the top and curl your shoulders. Curl up until your forearms touch your biceps.

TIMING CONSIDERATIONS

This exercise is effective because it offers continuous tension and an emphasis on the loaded stretch. It may be wise to use it at the end of a challenging training phase or during a biceps specialization block when you require a maximum stimulus.

SAFETY

Descend slowly and pause at the bottom. When the highest forces occur during the stretch, take care not to move explosively in that portion of the range of motion.

VARIATIONS

It's easy to rest between sets because you can return to the cable support by simply walking back and leaning over without even letting go of the handles. This makes this exercise ideal for myo-reps, so have fun! For extra variation, play around with cable heights and widths. See if you can find a few angle combinations that make you feel a high stimulus while keeping your joints comfortable.

CABLE "JERRY" CURL

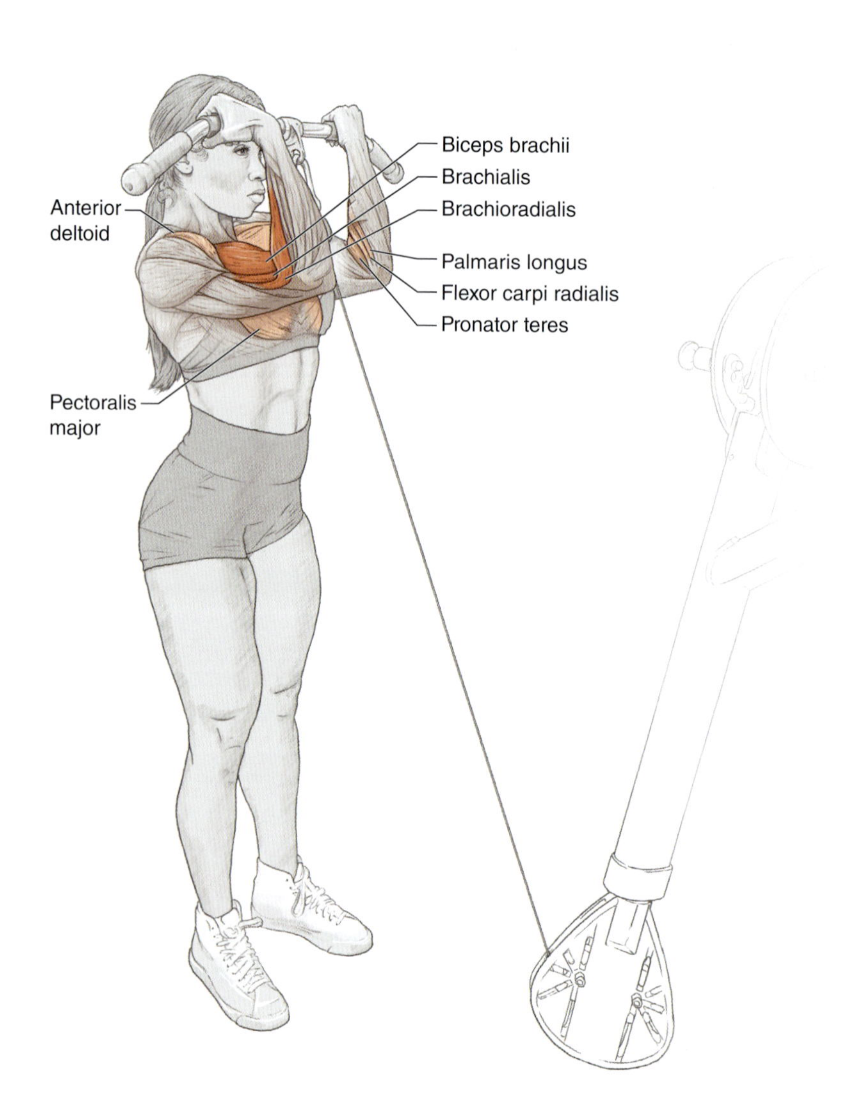

EXECUTION

1. Named jokingly after its biggest promulgator, Jared "Jerry" Feather (co-author of this book), this exercise begins like a regular cable curl but with a closer-than-normal grip. Ideally, your elbows should be positioned on top of your abdomen, just above your hip bones. Attach a pull-down bar to a low pulley on the cable machine. Stand facing the cable machine with feet shoulder width apart and grabbing the pull-down bar with a close grip.

2. Keeping your elbows flush to your abdomen just inside and above your hips, lean back and push your chest up before the movement begins. This will generate a lot of stretch in your biceps.
3. Once you begin to curl, focus on pushing your elbows together as much as possible.
4. Once your forearms touch your biceps at the top, return under strict control. Keep your elbows in enough so they again touch your abdomen (not just outside of it) at the bottom.
5. As you curl down on the eccentric phase, keep your chest high so that your biceps have to stretch to be able to get your elbows back enough to touch your abdomen at the bottom.

MUSCLES INVOLVED

Primary: Biceps brachii, brachialis, brachioradialis

Secondary: Pronator teres, flexor carpi radialis, palmaris longus, anterior deltoid, pectoralis major, pectoralis minor

EXERCISE NOTES

Start with a big chest and elbows in so that your biceps stretch the most possible at the bottom. If this means you can't touch your abdomen at the bottom, that's fine—as long as you feel a big stretch and are actively pulling your elbows together as much as possible during the movement. You can develop the skill to keep the elbows in by actively cueing to try to touch your elbows together the whole time, as if you're holding a ball between them and cannot let the ball drop. As you descend, resist the temptation to drop your chest. Keep it sky-high; you may even lean back a little so that maximum biceps stretch occurs.

TIMING CONSIDERATIONS

This exercise requires a lot of mental focus, so you shouldn't do it late in a session when you're mentally tired. Beginners might struggle with this advanced technique, so this exercise is probably best reserved for intermediates and above.

SAFETY

This exercise is very safe because it is low in loading potential when done properly. However, pushing your elbows together can result in cramping of the pecs on occasion; it may not be ideal to perform this exercise the day before a chest-focused workout.

VARIATIONS

You can perform this exercise with bars and dumbbells as well as cables.

ONE-ARM MACHINE PREACHER CURL

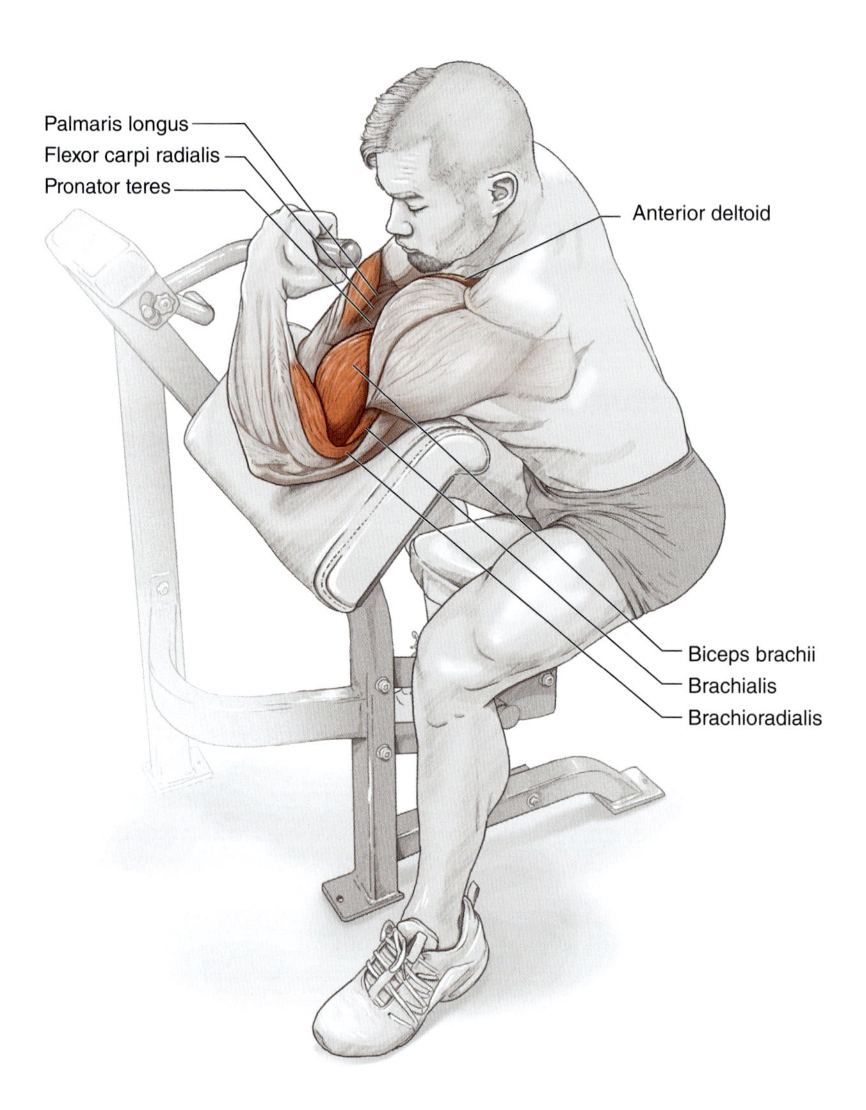

EXECUTION

1. At the preacher curl machine, set the bench height to where you will get a full stretch in the biceps (with your elbow in lockout) at the bottom of the curl. Sit at the machine and position one arm so your palm faces up instead of in as much as possible. The elbow joint should be directly under your forearm and not off to the side.
2. Feel a full stretch in the biceps at the bottom. Curl up all the way, trying to get your forearm to touch your biceps. Focus on keeping your palm in maximum supination by aiming your pinkie finger to the sky.
3. Return slowly on the eccentric phase, and repeat. Switch arms at the end of each set.

MUSCLES INVOLVED

Primary: Biceps brachii, brachialis, brachioradialis

Secondary: Pronator teres, flexor carpi radialis, palmaris longus, anterior deltoid

EXERCISE NOTES

Don't face forward in the preacher curl machine and just do one arm at a time, the way you might sit if you were curling both arms at the same time. Instead, sit facing a bit away from the arm that's curling so that you can get more supination and exposure to the biceps. To enhance the effectiveness, control the eccentric phase, and pause at both the top and bottom of the movement; holding at the top and even squeezing there for a second can allow you to develop a better mind–muscle connection with your biceps and get a lot more out of the eccentric phase.

TIMING CONSIDERATIONS

It will take more time to do this exercise since it's unilateral. Save it for biceps specialization phases or for when you have lots of time to commit to the gym.

SAFETY

It is unwise to bounce during this exercise. While it is rare to tear the biceps during curls (especially on machines), bouncing out of the hole with a high load increases your risk of injury.

VARIATIONS

Feel free to adjust the machine bench settings to find a comfortable position for the best feel in the biceps. You may also want to play with the seat height or curl bench height, since they might be independently alterable on some machines.

ONE-ARM BENCH WRIST CURL

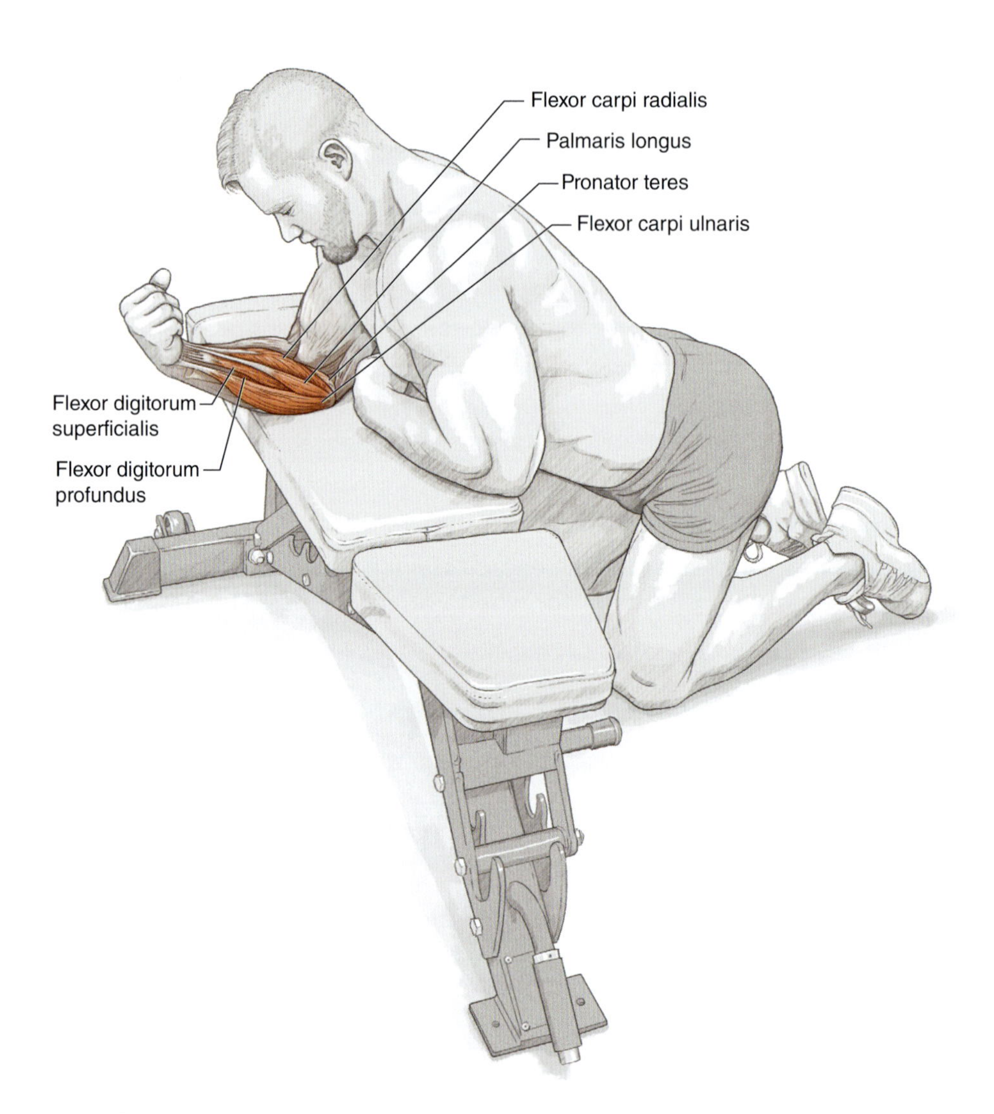

EXECUTION

1. Grab a dumbbell, kneel next to a flat bench, and lay your forearm across the bench with your palm up, dumbbell in hand. The dumbbell should be just over the edge of the bench so you don't nick it on the bench as you move it during the exercise.
2. Control the dumbbell all the way down until your wrist is fully extended, then flex your wrist up to above parallel.
3. Start the next rep in a slow eccentric movement. Switch hands at the end of the set.

NOTE: The dumbbell has been left off the illustration to better show the wrist muscles.

MUSCLES INVOLVED

Primary: Flexor carpi radialis, palmaris longus, flexor carpi ulnaris, flexor digitorum superficialis, flexor digitorum profundus

Secondary: Pronator teres

EXERCISE NOTES

Rushing this movement uses a lot of tendon rebound force and robs the muscles of their best growth stimulus potential. Slow down the eccentric phase, and gently restart the concentric phase at the bottom. The lower you can go without the dumbbell slipping out of your hands, the better; maximum stretch under load will lead to maximum results. On the concentric phase, you can either come up to just above parallel or to a fully flexed wrist, depending on which option gives you the best mind–muscle connection while avoiding wrist joint discomfort.

TIMING CONSIDERATIONS

Don't do this exercise before back or biceps training because your grip will become the limiting factor in back work and your forearms will become the limiting factor in biceps work. You can do this if you have straps or Versa Gripps, but even the sensation of a weaker grip can interfere with back and biceps training results.

SAFETY

Feel out your wrist position over the bench for comfort. You can place your wrist closer to the bench's edge or further away—whichever gets you to clear the bench while minimizing wrist discomfort.

VARIATIONS

Instead of placing your forearm across the bench width-wise, try placing your entire arm across the bench length-wise. The movement is identical except that your upper and lower arms will form a rigid segment, and your elbow joint will be locked. This variation can better target some of the wrist flexors that enter active insufficiency when the elbow is bent (such as the pronator teres, flexor carpi radialis, and palmaris longus muscles).

BARBELL STANDING WRIST CURL

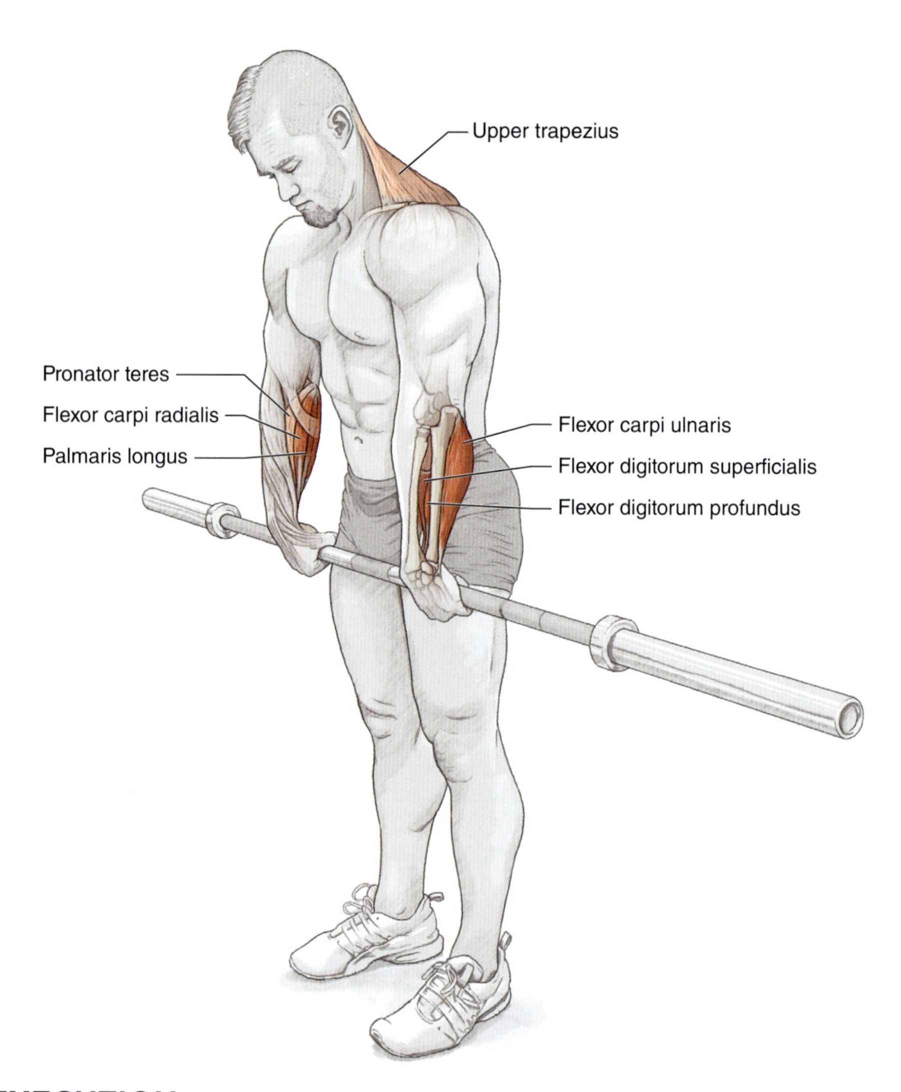

EXECUTION

1. Grab a barbell in a pronated, shoulder-width grip and stand straight up.
2. Curl only your wrists until no more movement is possible in the joint.
3. Hold that peak position for one full second. Slowly lower the bar to just barely in your grasp.

MUSCLES INVOLVED

Primary: Flexor carpi radialis, palmaris longus, flexor carpi ulnaris, flexor digitorum superficialis, flexor digitorum profundus

Secondary: Pronator teres, upper trapezius

EXERCISE NOTES

The peak contraction and one second hold at the top is a great idea for tracking, since it doesn't let you count the "almost reps" that can occur closer to failure if you're just momentarily getting to that top position. It also lets you squeeze your forearm muscles and form a huge mind–muscle connection that you can milk on the way down. While not mandatory, you can help increase friction by using chalk on this exercise; this allows your forearms to work harder without bar-slip holding them back. Lastly, resist shrugging or pulsing the weight up with your legs when the reps get tough; this adds needless systemic fatigue and makes it difficult to performance and progress on this exercise.

TIMING CONSIDERATIONS

While the loads used are small in the grand scheme, this exercise is performed while standing; this does add some axial fatigue. If your program already has loads of axially fatiguing moves (such as rows, deadlifts, good mornings, and squats), it may be wise to choose another forearm curl variation.

SAFETY

Your wrist joints should feel good when you're doing this movement. If they don't feel great, consider experimenting with a narrower or wider grip.

VARIATIONS

Varying the grip width is an easy way to expand your use of this exercise. Consider doing narrow grip forearm curls for a few months and then switching to wide grip forearm curls for the next few months.

FOREARM PUSH-OFF

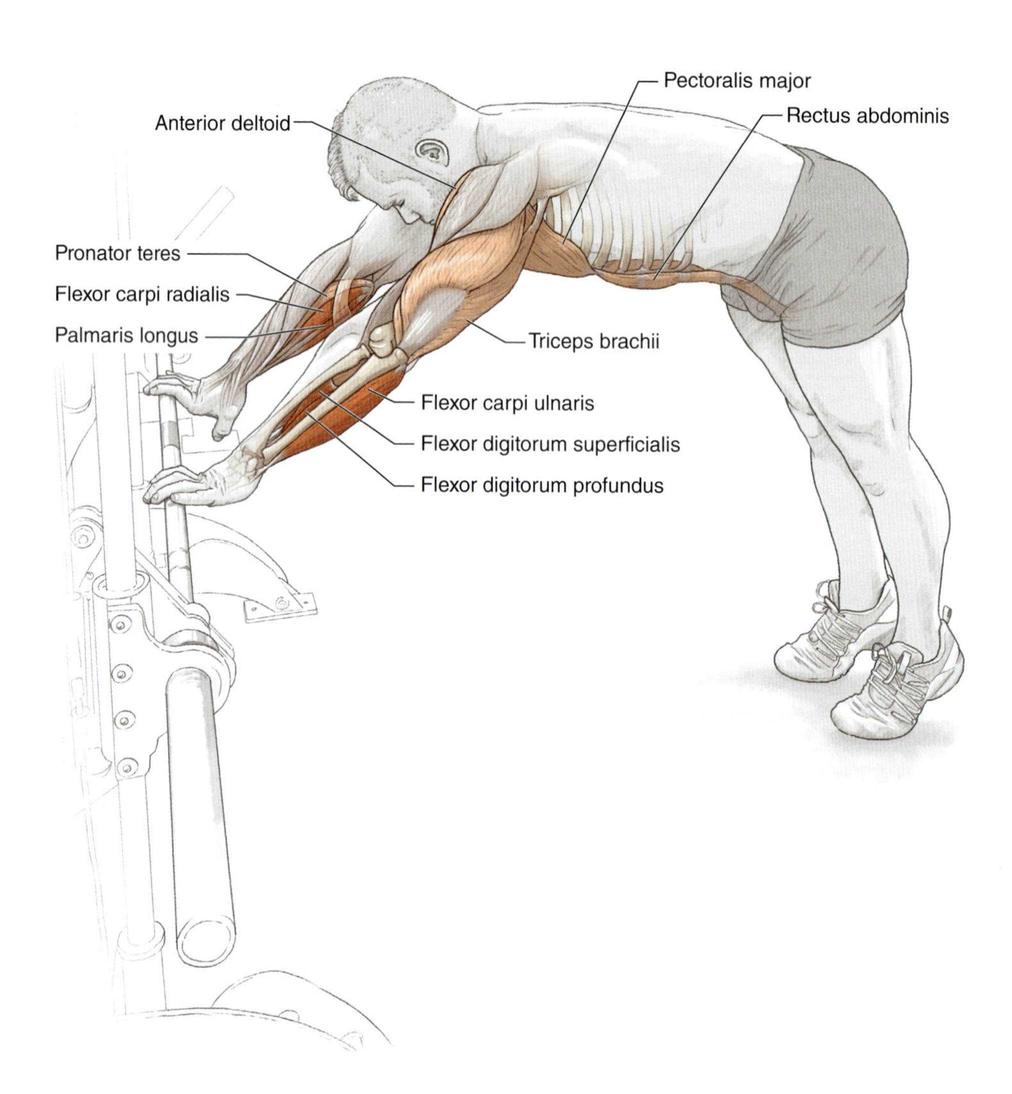

EXECUTION

1. Place a Smith machine bar at hip height or place a regular bar onto a power rack at hip height.
2. Step three to four feet (one meter) away from the bar. Reach forward and place your fingertips on the bar, keeping your entire upper and lower body as one rigid segment.
3. Extend your wrists and fingers to lower your body until your wrists and fingers are maximally extended.
4. Return to the starting position by flexing your wrists and fingers until your palms are about 15 degrees from parallel to your lower arms.

MUSCLES INVOLVED

Primary: Flexor carpi radialis, palmaris longus, flexor carpi ulnaris, flexor digitorum superficialis, flexor digitorum profundus

Secondary: Pronator teres, anterior deltoid, pectoralis major, pectoralis minor, rectus abdominis, transversus abdominis, internal oblique, external oblique, triceps brachii

EXERCISE NOTES

During the exercise, the lever arm elongates as you descend further into the eccentric contraction, maximizing the load during the stretch. Take your time in the eccentric phase (about two to three seconds total) and milk out the bottom, with a painful pause for about one or two seconds. You may be tempted to use momentum to come back up, especially by jutting your hips back and leaning away from the motion. Try to lean in instead, to make the exercise harder on your forearms, not easier.

TIMING CONSIDERATIONS

Because this exercise exposes your forearm musculature to a massive stretch under load, save this exercise for later mesocycles in a block, a macrocycle when you need the most stimulus, or forearm specialization phases.

SAFETY

If you go to muscular failure on this exercise, get ready to step forward to prevent falling down face-first!

VARIATIONS

Play with rotating bars or Smith bars to see what feels best. Rotating bars allow the movement to be smoother; however, sometimes they end up reducing the friction you need to get the best purchase with your grip.

DIP

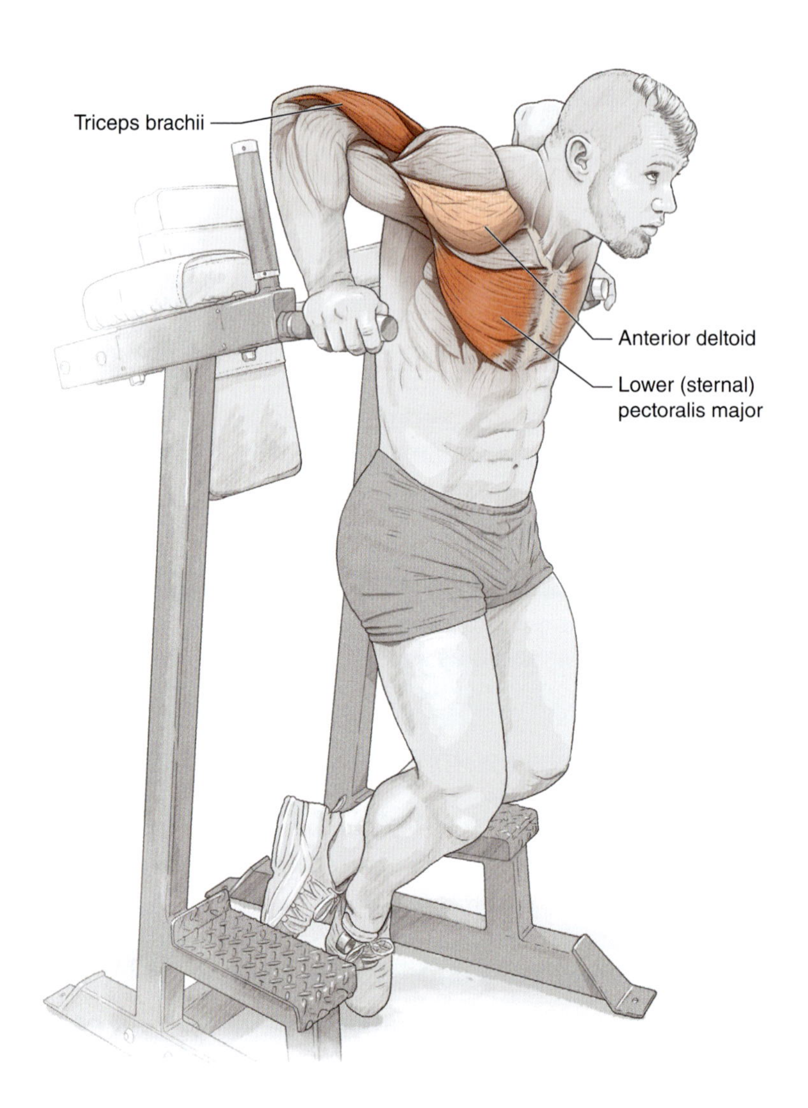

EXECUTION

1. Grab the parallel bars, which should be spaced shoulder-width apart. Hoist yourself up, with your arms straight and your elbows locked. While keeping your chest pointed up high and your feet either pointed down or tucked under and behind you, slowly dip down by pushing your elbows back. Continue to push them back as long as you're still descending.
2. Once you can no longer go down any farther, gently accelerate back up until the elbows are locked out. Then begin again.

MUSCLES INVOLVED

Primary: Triceps brachii, lower (sternal) pectoralis major

Secondary: Anterior deltoid

EXERCISE NOTES

Dips are very effective but can be a bit risky for the shoulder joints and pectoral muscles; always practice a slow, controlled descent. The bonus of such a descent is that it will allow you more time to assert your best technique and will greatly stimulate the triceps for growth. As you descend, you may be tempted to lean forward and use your front delts and pecs for most of the effort. Resist this temptation by pushing your elbows back as you descend. This increase in elbow angle directly stimulates the triceps more, especially in the deep stretch at the bottom. This means more growth, but your rep strength with this technique will be much lower than with the traditional technique, so expect the change. To help cue pushing the elbows back, try to keep your chest up.

TIMING CONSIDERATIONS

Dips are a great exercise for those light or strong enough to be able to do at least five (but hopefully 10 or more) reps. If that's not you, it is fine to do assisted dips or other triceps exercises.

SAFETY

There are two ways to enhance safety in the dip exercise. First, descend slowly and pause at the bottom before pushing up. This minimizes force spikes and reduces injury risk. Second, do not go lower than your shoulder comfort allows. If you feel consistent pain in the shoulder joints from dips, either stop short of the depth that gives you pain or switch to another exercise. The majority of people will be able to dip as low as they can with no shoulder issues.

VARIATIONS

The biggest factor in making dips an effective exercise for you is to find the hand position you like best. If you have a V-shaped dip bar, play around with narrower and wider grips to see where you seem to get the best stimulus for your triceps. The best stimulus will be the one that puts most tension on the triceps, gives them the biggest stretch, and pumps them up the most.

CABLE PUSH-DOWN

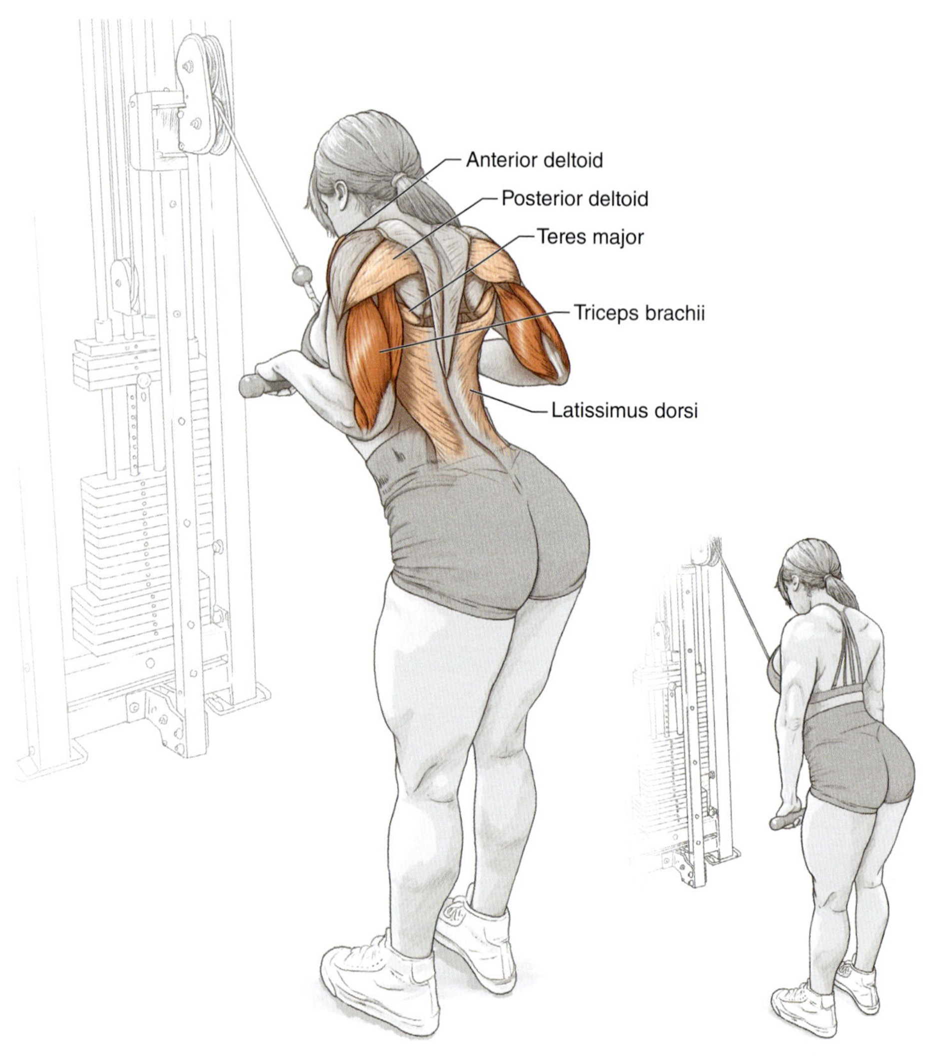

Push down.

EXECUTION

1. Place a cable attachment with a bar of your choice (a straight bar is a great place to start) on the cable machine at about eye level.
2. Stand very close to the bar, and then put an overhand grip on it. Pull the bar down so that the tops of your hands come as close as possible to touching your front delts.
3. While keeping your elbows back and in, push the cable down until your elbows are locked out.
4. On the ascent, push your elbows in and back, generating maximum tension under stretch in the triceps.

MUSCLES INVOLVED

Primary: Triceps brachii

Secondary: Posterior and anterior deltoid, teres major, latissimus dorsi

EXERCISE NOTES

Keep your chest up at all times. This will require the lower arm to travel further up to get to your front delt, enhancing the range of motion (especially at the all-important loaded stretch). This will also prevent your chest from being involved in most of the force production and support its recovery and adaptation from its own training. By pulling the elbows back on the way up, you will be able to maximize the loaded stretch on the triceps. You can also reduce the dynamic stabilization task required for the lats and teres major muscles, which will reduce their fatigue and enhance the results of their own training. While keeping your chest up and letting your elbows drift back on the way up, slow down the eccentric phase to maximize gains while doing the movement slowly enough so that you have the time to correct movement errors; this will improve your technique as well.

TIMING CONSIDERATIONS

The cable push-down is best for intermediate and advanced lifters. Beginners should focus on movements using free weights to build their general movement competency.

Because of the selectorized weight stack and your proximity to it, the cable push-down is ideally situated for drop sets if you choose to incorporate them.

SAFETY

Place your hands where your wrists, elbows, and shoulders feel best. There's no wrong answer here as long as your joints feel great and your triceps feel anything but!

JM PRESS

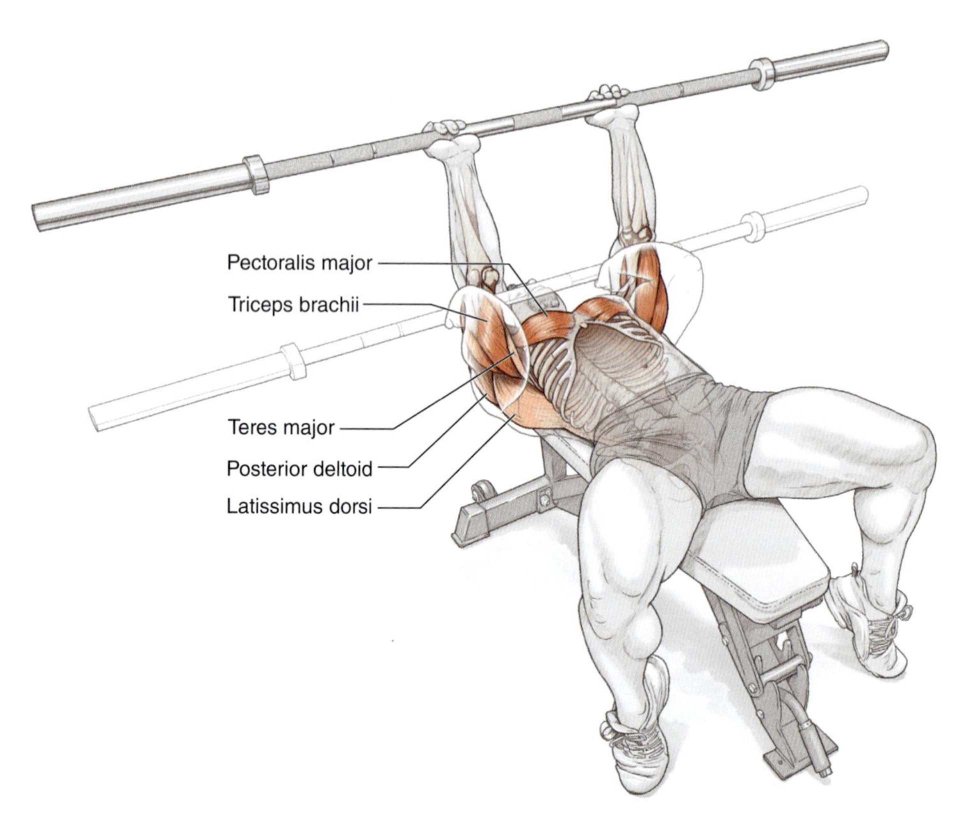

EXECUTION

1. Lie supine on a flat bench, with your feet flat on the floor. If your feet do not reach the floor, you may keep your toes planted on the floor, or you may consider putting a weight plate underneath your feet. Grip the barbell with your thumbs wrapped around the bar, and unrack the barbell. Begin with the barbell extended fully in a close-grip bench press body position and hand spacing (just at or inside shoulder width). Advanced lifters may find a false grip more comfortable.
2. Break at your elbows, and move them slowly toward your hips while keeping them from flaring out.
3. Continue the descent, and aim to have the tops of your hands end as close as possible to touching your anterior deltoids.
4. When you get as deep as your elbow flexion allows, return the bar to its initial position by extending up.

MUSCLES INVOLVED

Primary: Triceps brachii, pectoralis major

Secondary: Posterior deltoid, teres major, latissimus dorsi

EXERCISE NOTES

If you flare your elbows, your chest will contribute more to the movement. This isn't necessarily a bad thing, but to isolate the triceps, you'll want to keep your elbows in as much as you can. By pushing your elbows down to your hips during the descent, you can ensure the triceps take most of the load and experience the most loaded stretch at the bottom position; this is very growth-promoting. A slow eccentric movement is safer and potentially even more effective for hypertrophy. In addition, slowing the eccentric can make mastering the technique easier.

TIMING CONSIDERATIONS

If you like JM presses because they are easier on your elbows than skull crushers, you might like them even better by really leaning into the "elbows toward hips" cue. One of the biggest differences between JM presses (especially as traditionally prescribed) and skull crushers is the movement of the elbows toward the hips instead of toward the head.

SAFETY

Don't rush the bottom portion of this exercise. You will be putting a lot of force through the triceps tendons, so slow control is best.

VARIATIONS

Try this exercise on a Smith machine. When using the Smith machine, you can really focus on the mind–muscle connection because some of the control of the movement will be accomplished by the machine's track.

BARBELL SKULL CRUSHER

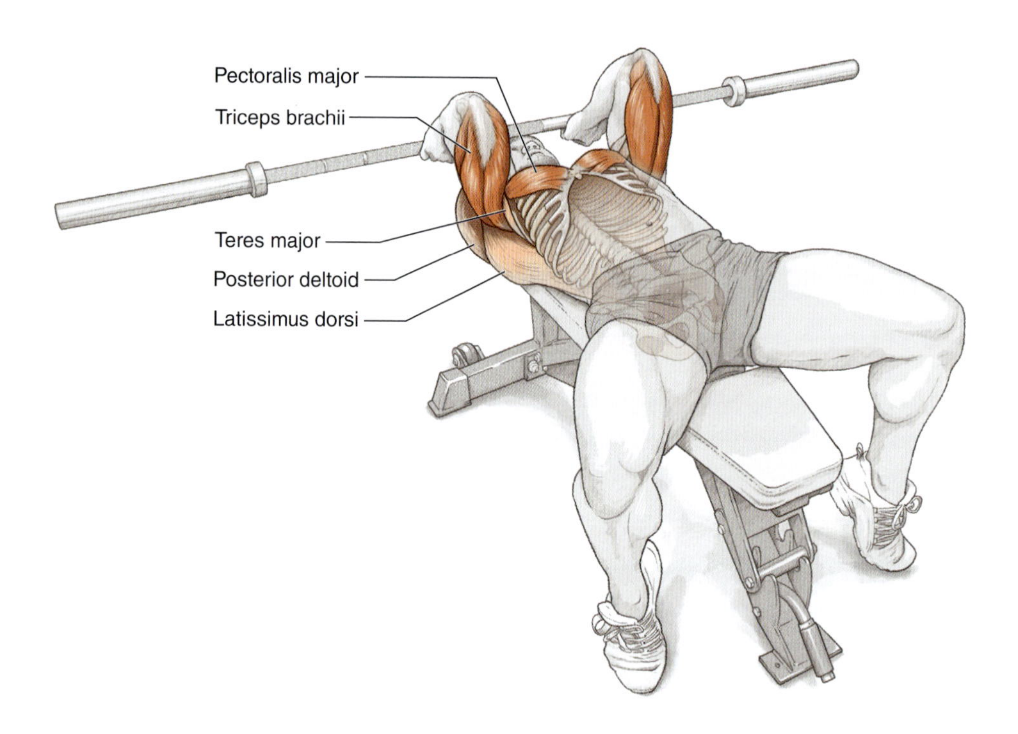

EXECUTION

1. Lie supine on a flat bench, with your feet flat on the floor. If your feet do not reach the floor, you may keep your toes planted on the floor, or you may consider putting a weight plate underneath your feet. Begin with the barbell extended fully in a close-grip bench press body position and hand spacing (just at or inside shoulder width).
2. Break at your elbows, and move them slowly toward your hips while keeping them from flaring out.
3. Continue the descent, and aim to have the bar touch the tip of your nose.
4. Once you touch your nose or go as deep as your elbow flexion will allow, return the bar to its initial position by extending up.

MUSCLES INVOLVED

Primary: Triceps brachii, pectoralis major

Secondary: Posterior deltoid, teres major, latissimus dorsi

EXERCISE NOTES

Flaring your elbows during this movement can increase the contribution of your chest, which may not always be undesirable. However, if you're looking to target your triceps, it is recommended that you try to keep your elbows in as much as possible. During the downward phase of the movement, push your elbows down toward your hips to ensure that your triceps bear most of the load and experience a significant stretch at the bottom position. To further optimize this exercise for hypertrophy, perform a slow eccentric phase. This can enhance your technique and safety while also potentially increasing the effectiveness of the exercise.

TIMING CONSIDERATIONS

A no-lose exercise in any phase, the skull crusher combines very well with supersets of close grip and especially with dumbbell work. Stand by the bench and grab a pair of dumbbells you can normally press for a set of 25 reps. Do as many skull crushers with the bar as you can, rack the bar, pick up the dumbbells, and then do as many flat dumbbell presses as you can. Your triceps will get bigger if they don't fall off!

SAFETY

Ease in and make sure that your elbows can handle this movement. Some people complain of elbow issues during skull crushers. Although the "elbows toward hips" cue often helps, starting with light weights and extended warm-ups can make a big difference in the tolerability of this exercise as well.

VARIATIONS

Try touching the bar to your neck, chin, eyes, forehead, and even the bench over your head. There's actually no correct way to perform either a skull crusher or a JM press. It's all a continuum of where you touch the bar, from roughly your clavicles to over your head.

INVERTED SKULL CRUSHER

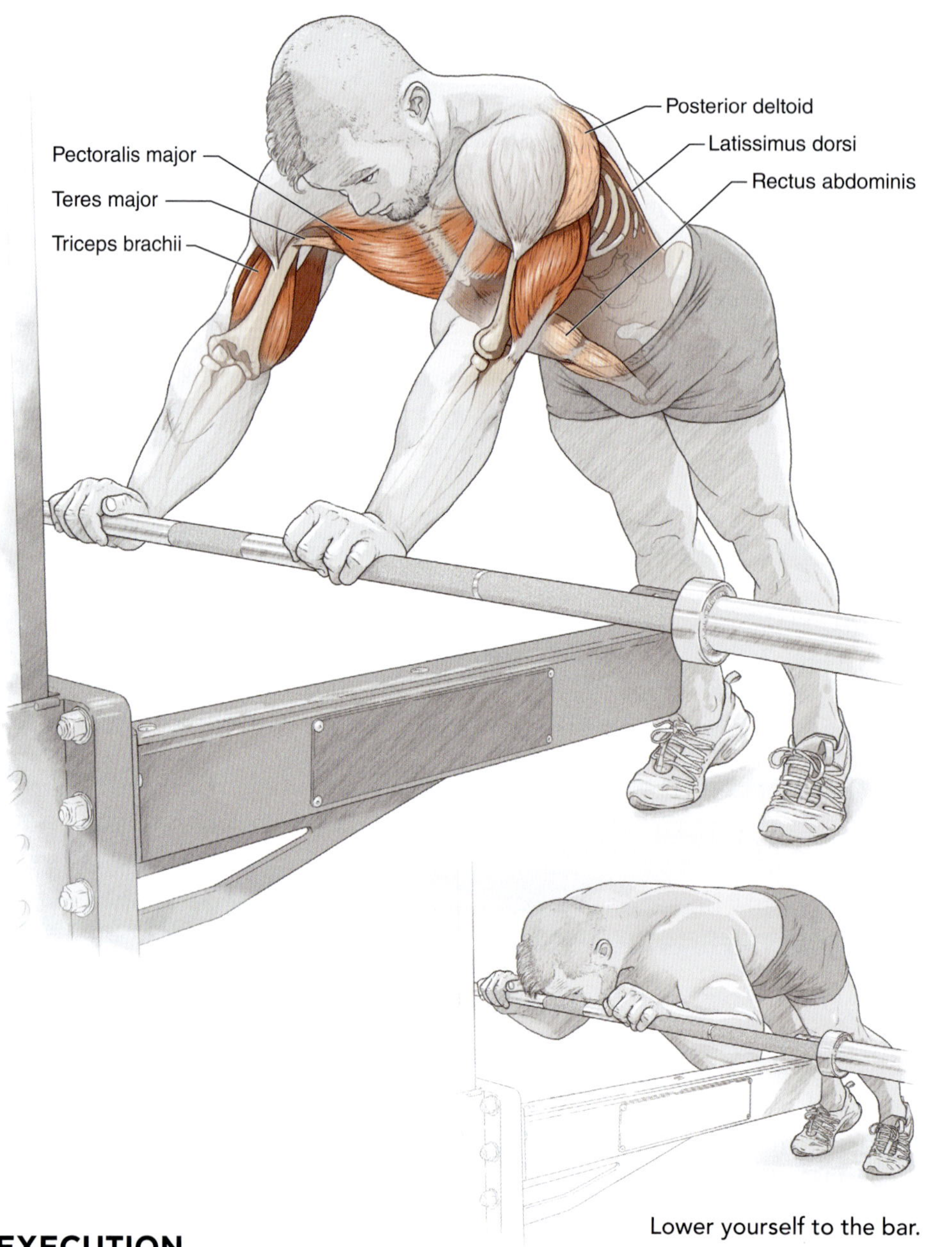

Lower yourself to the bar.

EXECUTION

1. Place a Smith machine or regular bar at about hip height. Stand back three to four feet (one meter) from the bar and lean over to grab it, hands just inside shoulder-width.
2. Keeping your upper and lower body as one rigid segment, break at your elbows and move them slowly toward your hips while keeping them from flaring out.

3. Continue the descent, and aim to have the bar touch the tip of your nose.
4. Once you touch your nose or go as deep as your elbow flexion will allow, return to your initial position by extending up.

MUSCLES INVOLVED

Primary: Triceps brachii, pectoralis major

Secondary: Posterior deltoid, teres major, latissimus dorsi, rectus abdominis, transversus abdominis, internal oblique, external oblique

EXERCISE NOTES

During the downward phase of the exercise, you can push your elbows toward your hips to help shift the emphasis onto your triceps and allow for a greater stretch at the bottom position. This can be beneficial for promoting muscle growth. Performing a slow eccentric phase can also help you execute the exercise with better technique and safety, potentially leading to better results for muscle hypertrophy. Flaring your elbows during the inverted skull crusher can increase the contribution of your chest muscles. However, you're likely to get your best results by tucking the elbows in as much as you can, especially on the descent.

TIMING CONSIDERATIONS

The heavier you are and the weaker you are, the higher up the bar should be placed to allow you to aim for any given repetition range with this exercise. As you get lighter and leaner over time, however, you might have to lower the bar to continue to keep this movement challenging.

SAFETY

When you're getting close to failure on this exercise, be prepared to tuck your knees up and step forward so that you don't end up getting a bit too intimate with the bar!

VARIATIONS

Just like in the skull crusher, you can try touching the bar to your neck, chin, eyes, forehead, and even reach up behind and let your head dip below the level of the bar (if that doesn't irritate your shoulder joints). You can aim for whichever point you'd like by walking a half step to or away from the bar in your starting position.

OVERHEAD EZ BAR EXTENSION

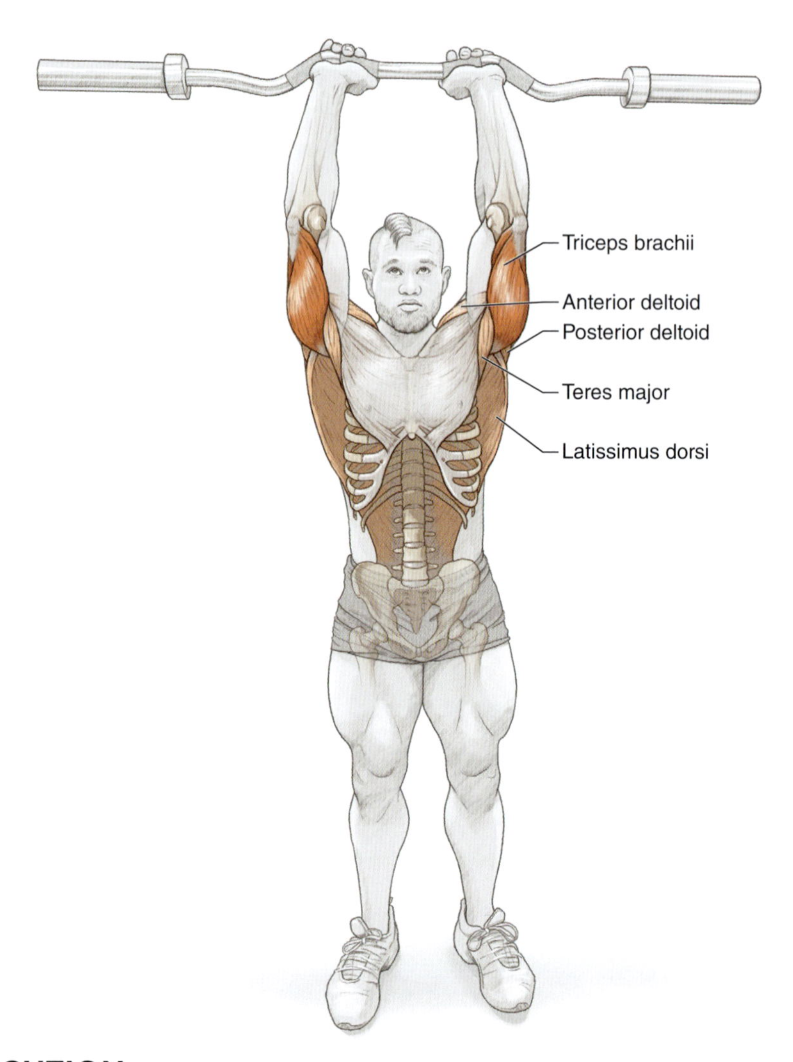

EXECUTION

1. Stand with your feet shoulder-width apart. Hold an EZ bar in an overhand grip, with your hands on either the inside or outside bends of the EZ bar.
2. Lift the bar over your head and extend your arms fully, with your elbows close to your ears.
3. Break at your elbows by pointing them straight ahead and pushing them forward in front of your face. This will lower the barbell behind your head; as it does, bend your neck forward a bit and let the barbell touch the lowest part of your neck that you can reach.
4. Pause for a moment at the bottom position, feeling the stretch in your triceps. Then gently accelerate the bar back up, moving your head back a bit once the bar clears it.

MUSCLES INVOLVED

Primary: Triceps brachii

Secondary: Posterior deltoid, anterior deltoid, teres major, latissimus dorsi

EXERCISE NOTES

People tend to reach very far back with this exercise, and then stop at a 90-degree angle of elbow flexion. This is a fine way to do this exercise, but going deeper is better for triceps stretch-mediated hypertrophy. Instead of reaching back as far as you can, tilt your head forward and reach your elbows as far forward as you can. Once your elbows are fully flexed at the bottom, then you can reach back and try to touch the back and bottom of your neck. The slower and deeper you go, the more triceps growth you will stimulate, especially if you keep your elbows in during the descent.

TIMING CONSIDERATIONS

This exercise will create some axial fatigue, so just keep that in mind.

SAFETY

While pausing at the bottom of the movement and not rushing the descent will decrease the injury risk of this exercise, in general, this exercise is already very safe. That being said, you can increase safety and avoid chronic wear and tear, especially for the shoulder and elbow joints, by taking the time to choose a hand position that leaves these joints most comfortable (or at least by choosing one of the many hand positions that doesn't actively irritate them).

VARIATIONS

Some basic variations include grabbing the bar on the inside of the bends, on the outside of the bends, and even on the bends of the EZ bar itself.

OVERHEAD CABLE EXTENSION

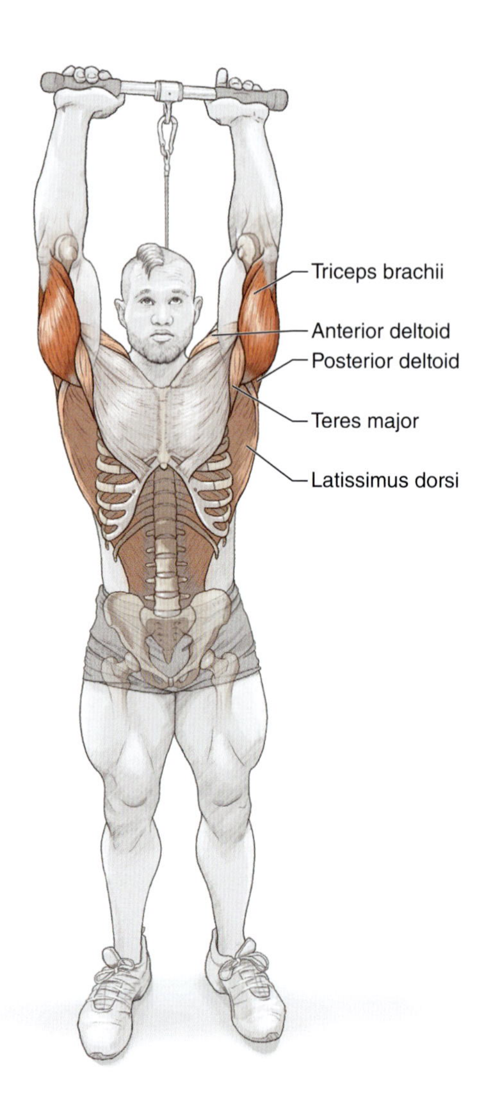

EXECUTION

1. Put either a straight or EZ cable attachment on the cable machine at just below shoulder height and duck under it. Facing away from the stack, press the cable attachment to full extension over your head, just about one foot (about one-third of a meter) in front of the cable stack itself.
2. Break at your elbows by pointing them straight ahead and pushing them forward in front of your face. This will lower the bar behind your head; as it does, bend your neck forward a bit, and let the bar touch the lowest part of your neck that you can reach.
3. Pause for a moment at the bottom position, feeling the stretch in your triceps. Then gently accelerate the bar back up, moving your head back a bit once the bar clears it.

MUSCLES INVOLVED

Primary: Triceps brachii

Secondary: Posterior deltoid, anterior deltoid, teres major, latissimus dorsi

EXERCISE NOTES

Because flaring the elbows forward might cause the cable to scratch your back and neck on this exercise, it pays to reach back during the descent. By going slow and keeping the elbows in, you can maximize the growth stimulus. It's fine to flare the elbows out a bit on the ascent; however, the more you can keep them pointing forward, especially on the descent, the more you'll engage the triceps and the less you'll ask of the anterior deltoids.

TIMING CONSIDERATIONS

Because this exercise is performed while standing, it will create some axial fatigue. This is something to keep tabs on if your training week already includes many other movements that create axial fatigue (such as overhead presses and deadlifts).

SAFETY

To reduce the risk of injury, take a brief pause at the bottom of the movement, and avoid a rapid descent. That said, this exercise is generally considered to be safe, and has a low risk of acute injury. To further minimize the risk of chronic wear and tear, particularly in the shoulder and elbow joints, choose a hand position that is comfortable and doesn't cause irritation.

VARIATIONS

There may be a higher number of different kinds of bars to use for cable overhead extensions than stars in the sky. If you have access to different bars, feel free to experiment!

ONE-ARM OVERHEAD DUMBBELL EXTENSION

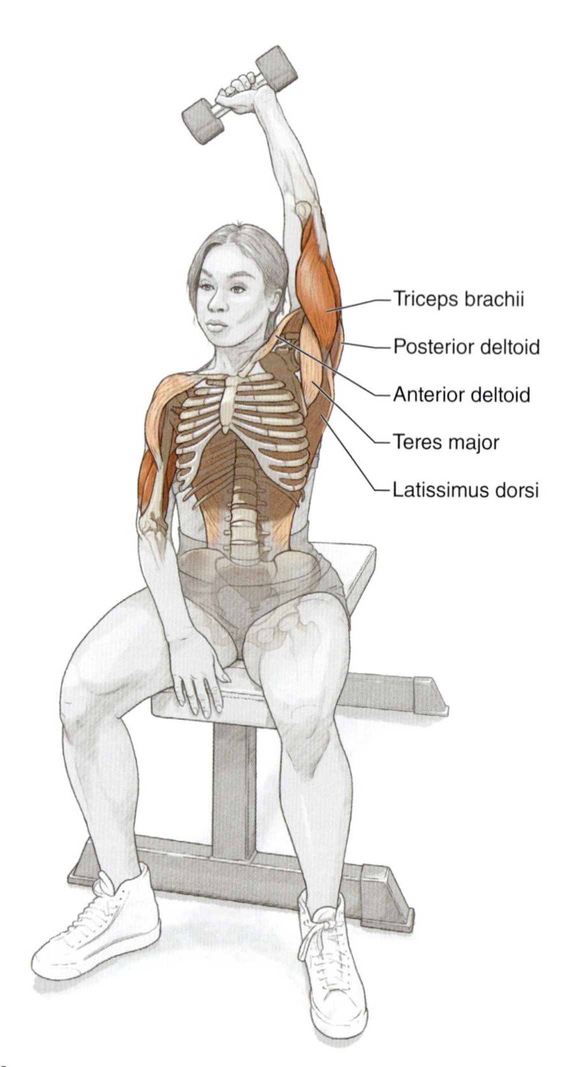

EXECUTION

1. Sit or stand upright, with your feet flat on the ground. If you stand, choose a stance that is comfortable for you (such as feet shoulder-width apart, wider than shoulder-width, or a staggered stance). Grab a dumbbell on the outermost edge of its handle so the outside of your fist is touching the inner weight plate.
2. Push press the dumbbell overhead, stabilize it, and then begin a slow descent toward the back of your neck (as if there's something itching there and you want to use the dumbbell to scratch it). Try to touch the dumbbell to your upper back, at the midpoint between your spine and your glenohumeral joint, just below your trapezius.
3. Once there, pause for a second, and extend the dumbbell back up. Switch arms after finishing all of the reps within the set.

MUSCLES INVOLVED

Primary: Triceps brachii

Secondary: Posterior deltoid, anterior deltoid, teres major, latissimus dorsi

EXERCISE NOTES

Because this exercise works one arm at a time, it will allow you to completely focus on the triceps and away from the anterior deltoid. To maximize triceps focus and especially target the long head of the triceps, aim to reach back as far behind you as you can. Where your elbow points during the descent is not really an issue, so long as your shoulder joint is comfortable. By going super slow on the way down and getting to a maximum bottom stretch, you can stimulate a considerable amount of triceps growth with this exercise (especially if you can extend back up on the concentric instead of pressing).

TIMING CONSIDERATIONS

If you choose to do this exercise standing upright, some axial fatigue is likely, so just keep that in mind. However, if you can sit down while doing this, you'll win on all counts!

SAFETY

This is a very safe exercise. However, because it's a dumbbell overhead movement, it's not overly stable. Make sure to take a wider or staggered stance. Consider either holding your non-extending hand on your hip or even holding onto a rack or railing to increase stability. The slower your descent is on this movement, the safer it will be.

VARIATIONS

You can do this exercise seated or standing. You can use a cable attachment instead of a dumbbell as well, so give that some thought. You can also experiment with different wrist orientations, from pointing your wrist up to the ceiling at the bottom to taking a neutral grip in which your pinkie finger points to the ceiling instead. Both variations are effective, but they target subtly different parts of your triceps; one or the other may be more comfortable for your shoulder or elbow joints.

CHAPTER 5

Abdominals

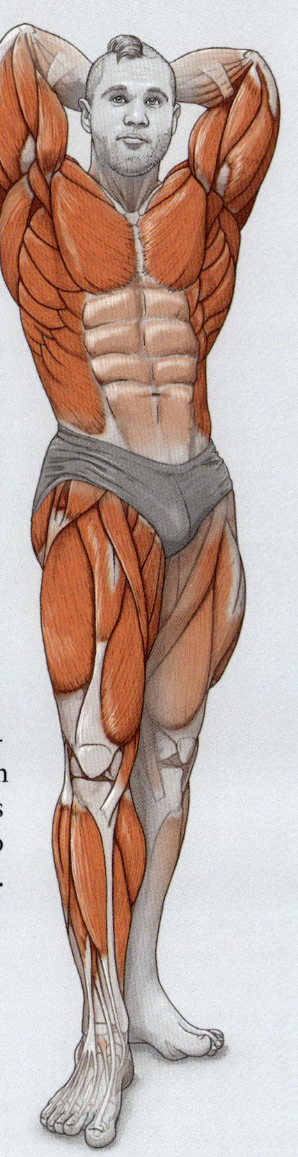

The abdominal muscles (figure 5.1), also known as the "abs," extend from various points on the ribs and spine to the pelvis and pubic bone. It is helpful to divide the group into two sections: the front wall and the side wall. The front abdominal wall consists of one muscle, the rectus abdominis. This is the muscle responsible for the highly sought-after "six pack." The rectus abdominis is a paired, segmented muscle that originates from the crest of the pelvis and inserts onto the xiphoid process of the sternum and the costal cartilages of ribs 5 to 7. The rectus abdominis muscles are separated at the midline by a band of connective tissue called the linea alba. The tendinous intersections divide the rectus abdominis horizontally into separate muscle bellies. The main function of this muscle is to flex the lumbar spine.

The side abdominal wall consists of three layers of muscle, from superficial to deep: the external oblique, internal oblique, and transversus abdominis. The external oblique muscle is a paired muscle located on each side of the rectus abdominis; it originates from ribs 5 to 12 and inserts onto the pelvis. Its function is to flex the torso and rotate the torso to the opposite side.

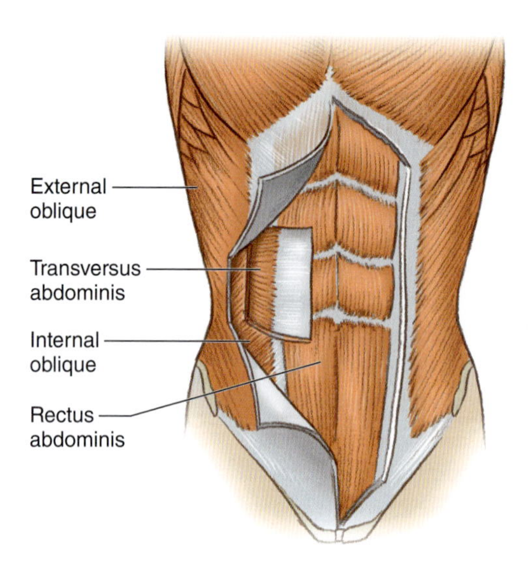

Figure 5.1 Abdominal muscles.

The internal obliques are located just underneath (deep to) the external obliques. They are found on each side of the rectus abdominis, running from the lower back around to the sides and front of the abdomen. They originate from the lumbar fascia and pelvis, insert into the inferior borders of ribs 10 to 12 and the linea alba. The internal obliques function to laterally flex the torso to the side and to rotate the torso to the same side.

The transversus abdominis lies underneath the internal oblique muscles. It is often referred to as the "corset muscle" since it wraps around the abdomen and spine, providing support and stability to the spine and pelvis. The transversus abdominis originates from the costal cartilages of ribs 7 to 12, the thoracolumbar fascia in the lower back, and the pelvis. It inserts at the linea alba and pelvis. In addition to providing stability and support, the transversus abdominis is responsible for compression of the abdominal contents. Most notably, when bodybuilders perform vacuum exercises to make the abdomen appear less bulgy, it is the transversus abdominis muscle that is responsible for making it happen.

CANDLESTICK

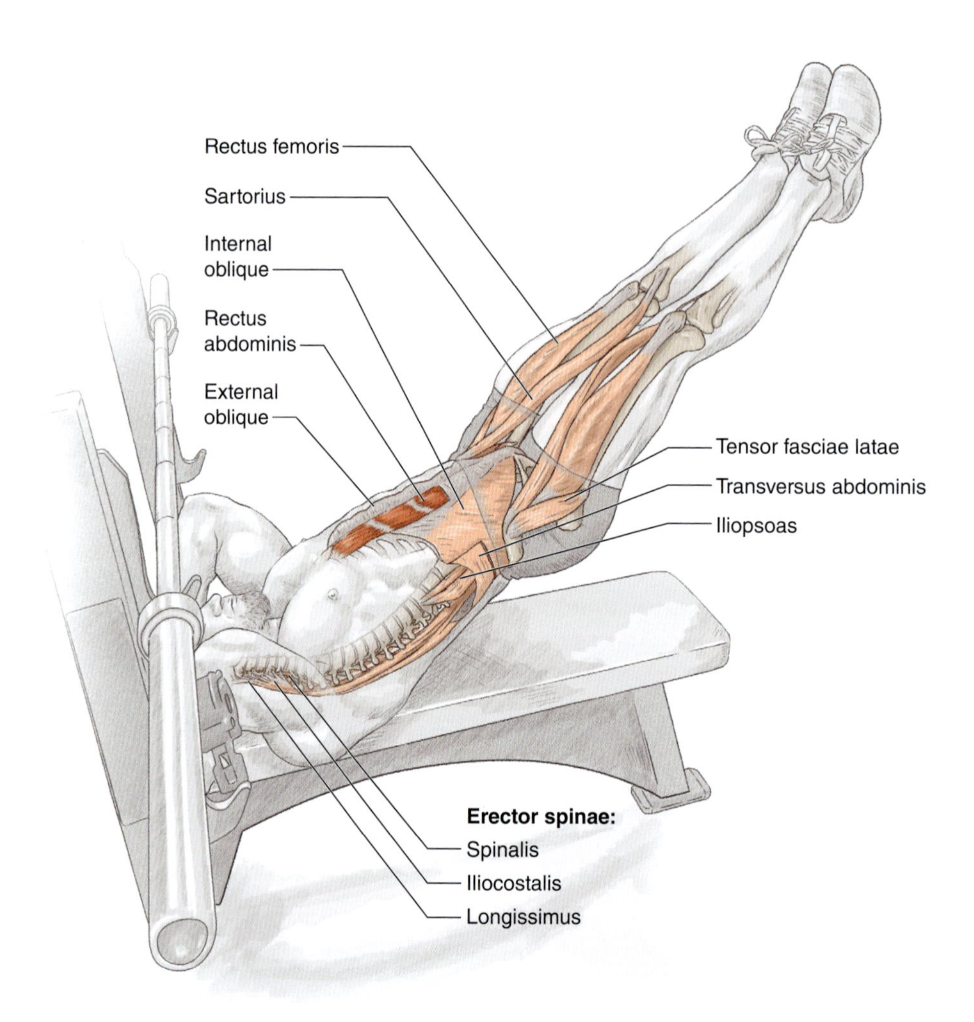

EXECUTION

1. Lie face up on a bench. Place your arms overhead behind you, and tightly grab the bench on both sides.
2. Keeping your legs completely straight, bend your hips until your legs are pointed vertically, with toes pointed vertically as well.
3. Contract your abs and curl your entire abdomen up, so that your hips lift completely up from the bench (as high up and toward your head as they can go). As you lift your hips, continue to point your legs and feet straight up.
4. Pause briefly at the top. Then reverse the movement slowly, first gently lowering your hips and then gently lowering your legs back down to parallel with the ground.

MUSCLES INVOLVED

Primary: Rectus abdominis

Secondary: Iliopsoas, rectus femoris, sartorius, tensor fasciae latae, external oblique, internal oblique, transversus abdominis, erector spinae (iliocostalis, longissimus, spinalis)

EXERCISE NOTES

Don't worry about ascent speed on this exercise. The real money on this exercise is made in the eccentric phase (as you will painfully and hilariously discover when you attempt it for the first time). To that end, make sure to really focus on doing the eccentric slowly and under control, with a high degree of mind–muscle connection to the abs. Rushing the eccentric renders this exercise somewhat pointless. Lastly, make sure to go all the way down to the bench with both your hips and legs. Cutting the range of motion is as tempting as it is counterproductive.

TIMING CONSIDERATIONS

This is a brutal and difficult exercise that requires a high baseline of strength in the abdominals and hip flexors. Most people who are beginning to train their abs should not start with this exercise and should instead opt for the cable crunch and abdominal machine exercises until they are much stronger. At that point, the modified candlestick (see variations) becomes an option.

SAFETY

Get a good grip on the bench for safety and to generate a lot of tension with your abs. The body prefers to generate its highest forces only when most stable, so make sure you feel really good about that grip.

VARIATIONS

If you are not as strong, you can bend your knees when bringing your legs up, extending them once your thighs are vertical, and then flexing at your torso. When you get stronger, perform this exercise with straight legs. The ultimate challenge is to do the unmodified candlestick: Keep your hips completely rigid from the start, and arc your toes to vertical simply by flexing at your core. If you can do a set of 10 of these, consider yourself in possession of truly rare, elite ab strength. If you are able to hold onto a stable object, you may perform this exercise directly on the floor.

HANGING LEG RAISE

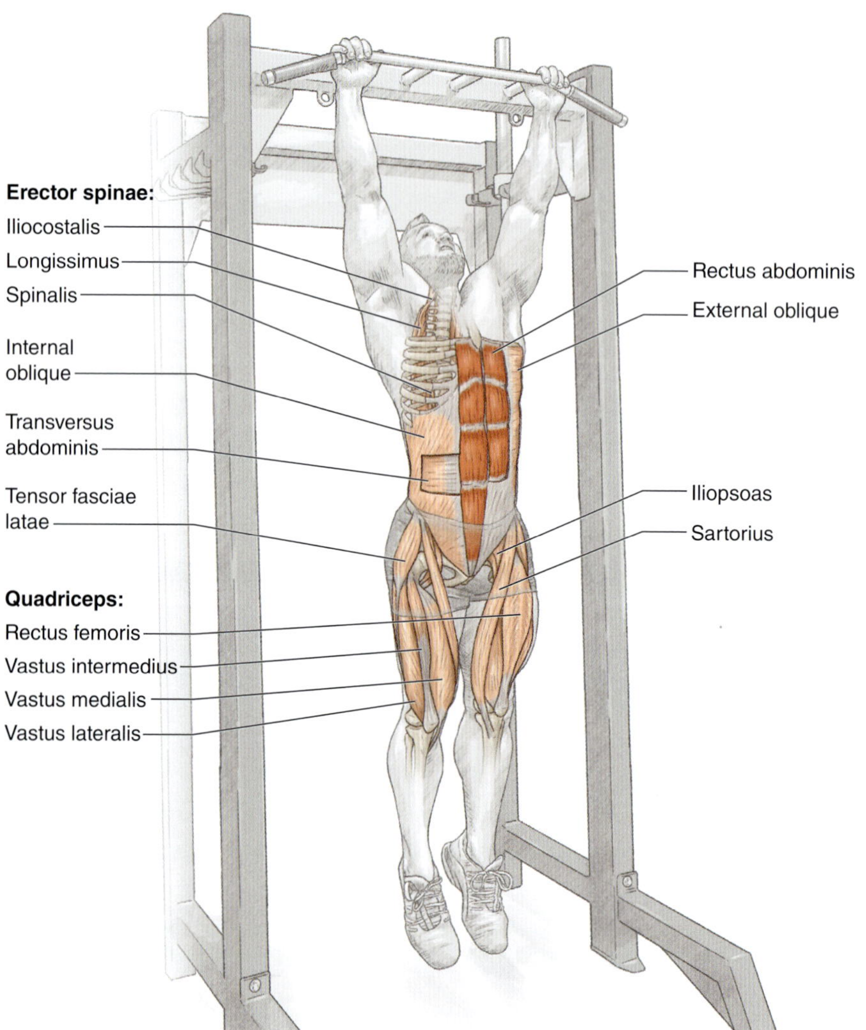

EXECUTION

1. Hang from a bar using a roughly shoulder-width grip, and let your legs hang down freely.
2. Keeping your knee joints straight and rigid, lift your feet (and your legs) in an arc out in front of you.
3. Lift your legs as high as you can; on an ideal rep, your feet should touch the bar you're hanging from.
4. Reverse the motion in the same arc, under slow eccentric control.

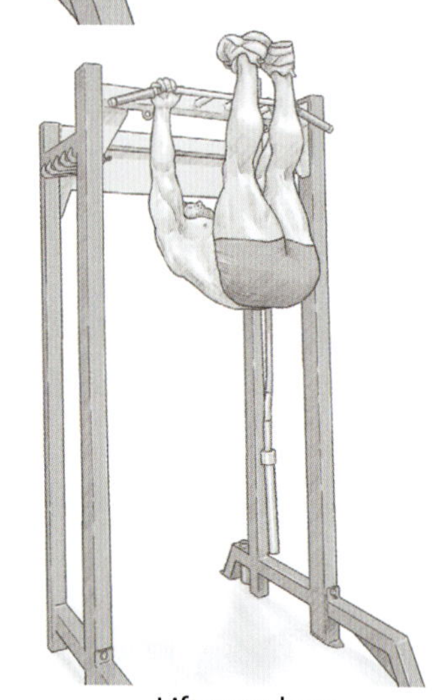

Lift your legs.

MUSCLES INVOLVED

Primary: Rectus abdominis

Secondary: Iliopsoas, rectus femoris, sartorius, tensor fasciae latae, external oblique, internal oblique, transversus abdominis, quadriceps (rectus femoris, vastus lateralis, vastus medialis, vastus intermedius), erector spinae (iliocostalis, longissimus, spinalis)

EXERCISE NOTES

Go as high as you can on the concentric phase, ideally touching your toes to the bar. If you can't do this, go as high as you can every time, at least getting your legs above parallel to the ground. With every rep, slow the eccentric phase, especially when your legs are close to parallel to the ground and the lift is at its hardest. It is easy to slow the eccentric phase at the bottom, when you're barely fighting gravity anymore; it is very tough (but confers a huge benefit) to slow the eccentric phase in the middle range or when your legs are parallel to the ground.

TIMING CONSIDERATIONS

This is a quick and simple exercise to perform, but you need a pull-up bar or something like it to hang from. Prepare in advance if you anticipate a busy gym with limited access to a pull-up bar. If you can't gain access to the bar, consider another exercise like the V-up that relies on open floor space and no implements.

SAFETY

Wearing straps or Versa Gripps can be a huge help, not just to make you confident about your grip, but also to allow you to focus on the abdominal work instead of thinking about whether your grip is slipping. If you wear these, stand on a box about a foot behind the pull-up bar as you put them around the bar. Then gently step off, and gently step back and up once you're done with a set so that you can disconnect yourself from the bar.

VARIATIONS

If you're not strong enough to get a lot of reps with straight legs, do this exercise with bent knees, driving your knees up to your chest and then slowly controlling their descent. If even that's a bit much at first, try just getting your thighs above parallel with bent knees and slowly controlling the descent. Once you are able to complete sets of at least 20 reps, try some reps with your knees coming all the way to your chest. Once you can do at least 20 reps of those, you can transition to the straight-legged version.

SLANT BOARD SIT-UP

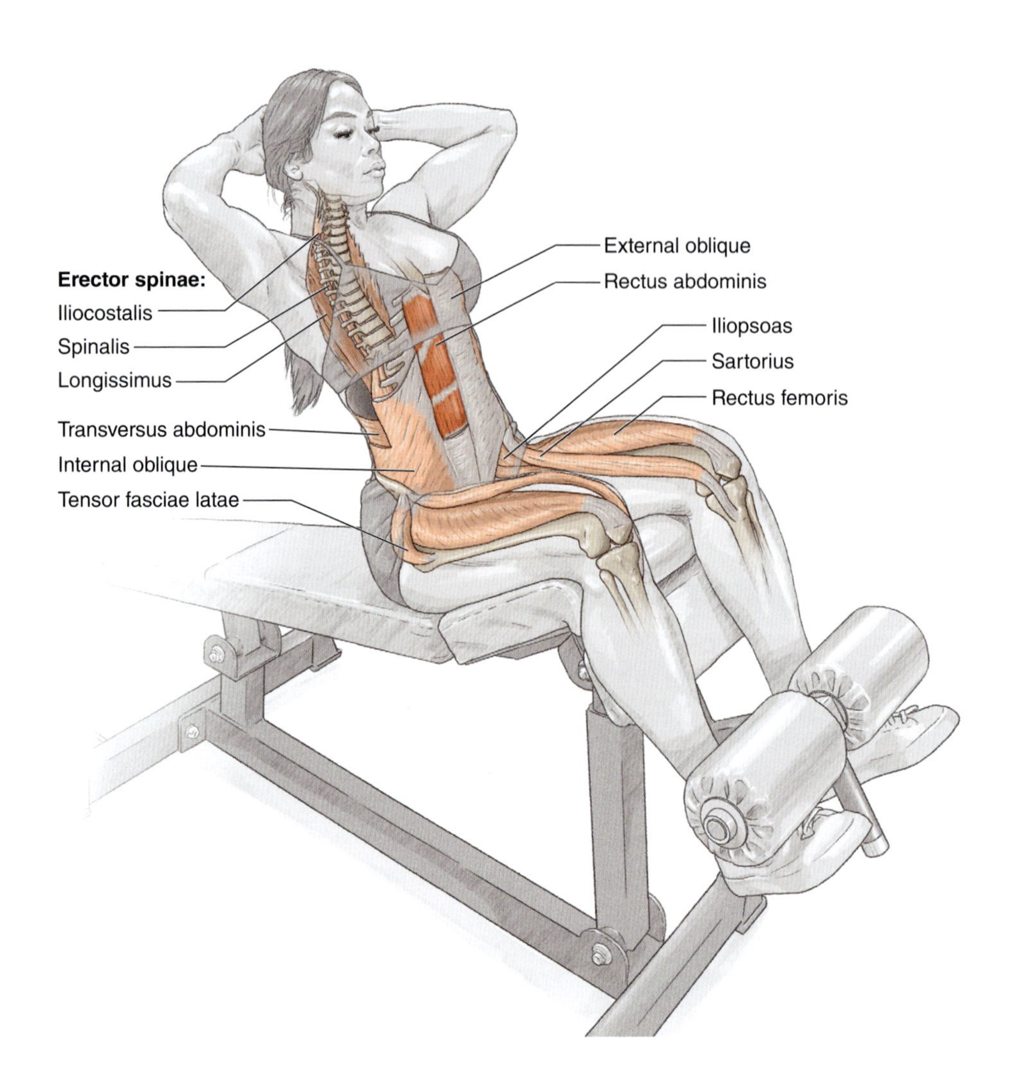

EXECUTION

1. Lie supine on a slant board. Tuck your feet firmly into the restraint. Grab a weight (dumbbell, kettlebell, or weight plate) and either hold it overhead or to your chest.
2. Curl your abdomen up until your chest is perpendicular to the ground.
3. Slowly return to the board under control.

MUSCLES INVOLVED

Primary: Rectus abdominis

Secondary: Iliopsoas, rectus femoris, sartorius, tensor fasciae latae, external oblique, internal oblique, transversus abdominis, erector spinae (iliocostalis, longissimus, spinalis)

EXERCISE NOTES

Hold the weight in the most comfortable way possible. You want to be able to focus on the abs instead of worrying about gripping the weight. Holding a dumbbell or plate across the chest is often a good move here. Focus on a slow eccentric phase, keeping constant tension on the abs, even at the bottom of the movement. This will make the exercise harder than if you relax at the bottom between reps, but harder is the goal!

TIMING CONSIDERATIONS

Because this exercise requires a slant board, it might be tough in a busy gym. If that's the case, you can opt for exercises with less specific equipment requirements such as the V-up.

SAFETY

If you keep your abs flexed and your spine just barely in flexion even at the bottom, you can avoid having to transition from spinal extension to flexion at the bottom. Some individuals report back pain from such a transition; while it's almost certainly not inherently unsafe, it's something to consider.

VARIATIONS

Where you hold the load can make a big difference in how tough this movement is. If you don't have a heavy weight or hate holding it to your chest, use a lighter weight and keep your arms over your head the whole time. This will magnify the difficulty of the exercise, especially at the bottom stretch where it's most stimulating to muscle growth.

V-UP

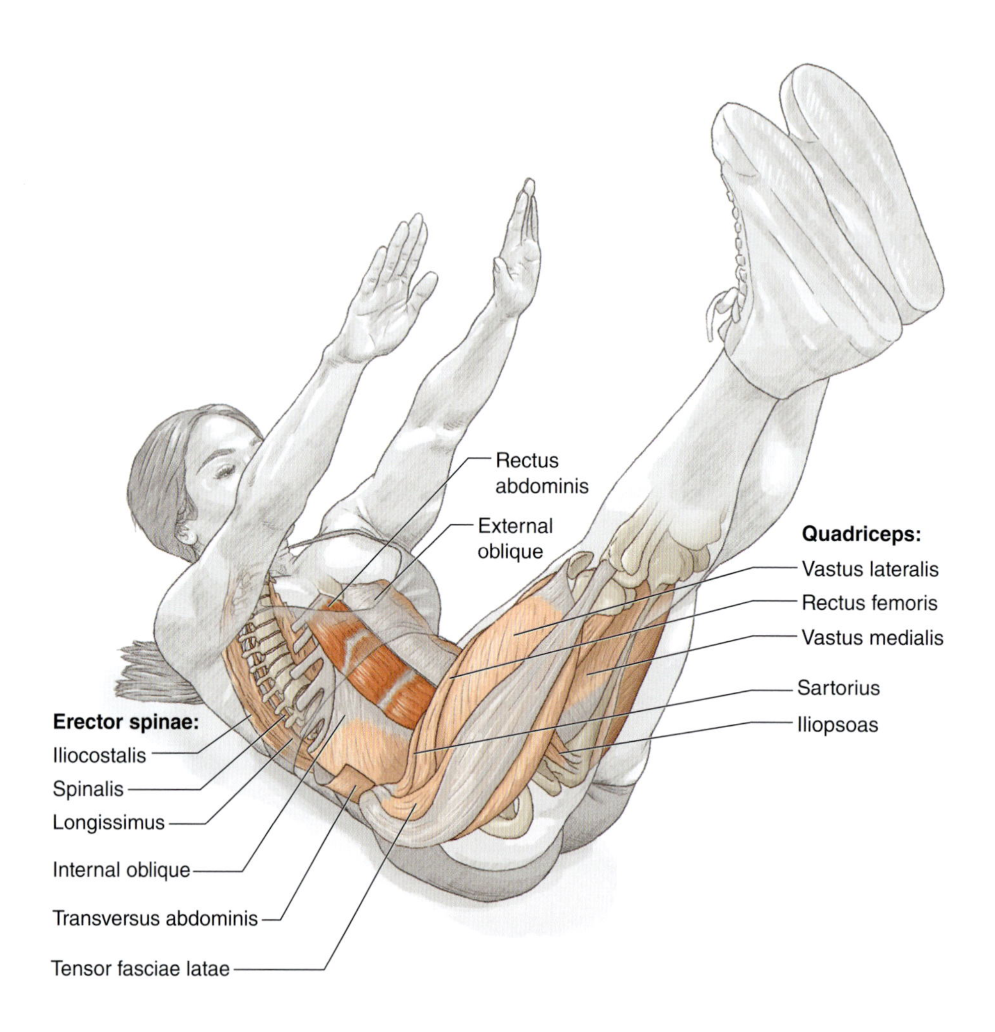

EXECUTION

1. Lie supine on the ground. Make your legs (across the knee joint) as rigid as possible. Stretch your arms above and behind you, make the elbow joints rigid, and touch the ground.
2. Holding a weight in both hands, simultaneously arc both your legs and your arms up and toward the middle of your body by flexing at the hips, extending at the shoulders, and, most importantly, flexing at your spine.
3. Once you gently touch either your toes or your shins to the weight in your hands, arc back down exactly as you came up—but slower and under strict eccentric control.

MUSCLES INVOLVED

Primary: Rectus abdominis

Secondary: Iliopsoas, rectus femoris, sartorius, tensor fasciae latae, external oblique, internal oblique, transversus abdominis, quadriceps (rectus femoris, vastus lateralis, vastus medialis, vastus intermedius), erector spinae (iliocostalis, longissimus, spinalis)

EXERCISE NOTES

Try to touch your hands (or the weight in your hands) to your feet at the top, and hold that position for a second. You can touch your knees or shins especially if your arms and torso are shorter, relative to your legs. However, reaching for your toes does the best job of promoting the arcing motion that so effectively engages your abs. Once you begin to descend from this position, descend slowly and with control, because that eccentric contraction is a huge driver of muscle growth. Focus! This exercise will suffer the most if you just go through the motions.

TIMING CONSIDERATIONS

Some people might find balance difficult with this exercise, because coordinating our hands and feet to touch at the same time can initially be a challenge. Take the movement slowly and minimize momentum-based swinging; this will help you balance and will make the exercise more challenging for the target muscles. If you're really strong, you can make this exercise even more challenging by holding the weight in your hands as far up and behind you as possible, only minimally extending at the shoulders to touch your toes at the top. This will both maximize stretch under load (and thus growth) and require you to sit up very high and forward at the top, blasting your abs.

SAFETY

At the bottom, don't flatten out and relax completely. Keep a slight curve to your spine by keeping your abs active between each rep. This will make the exercise tougher, more effective, and potentially even a bit safer.

VARIATIONS

If you're not super strong on this exercise when just starting out, see if you can work up over time to sets of at least 20 reps by bringing your feet and arms up to 45 degrees above horizontal. Once you're able to do that, you can begin to do the full movement, touching your hands to your feet or shins every time.

MACHINE CRUNCH

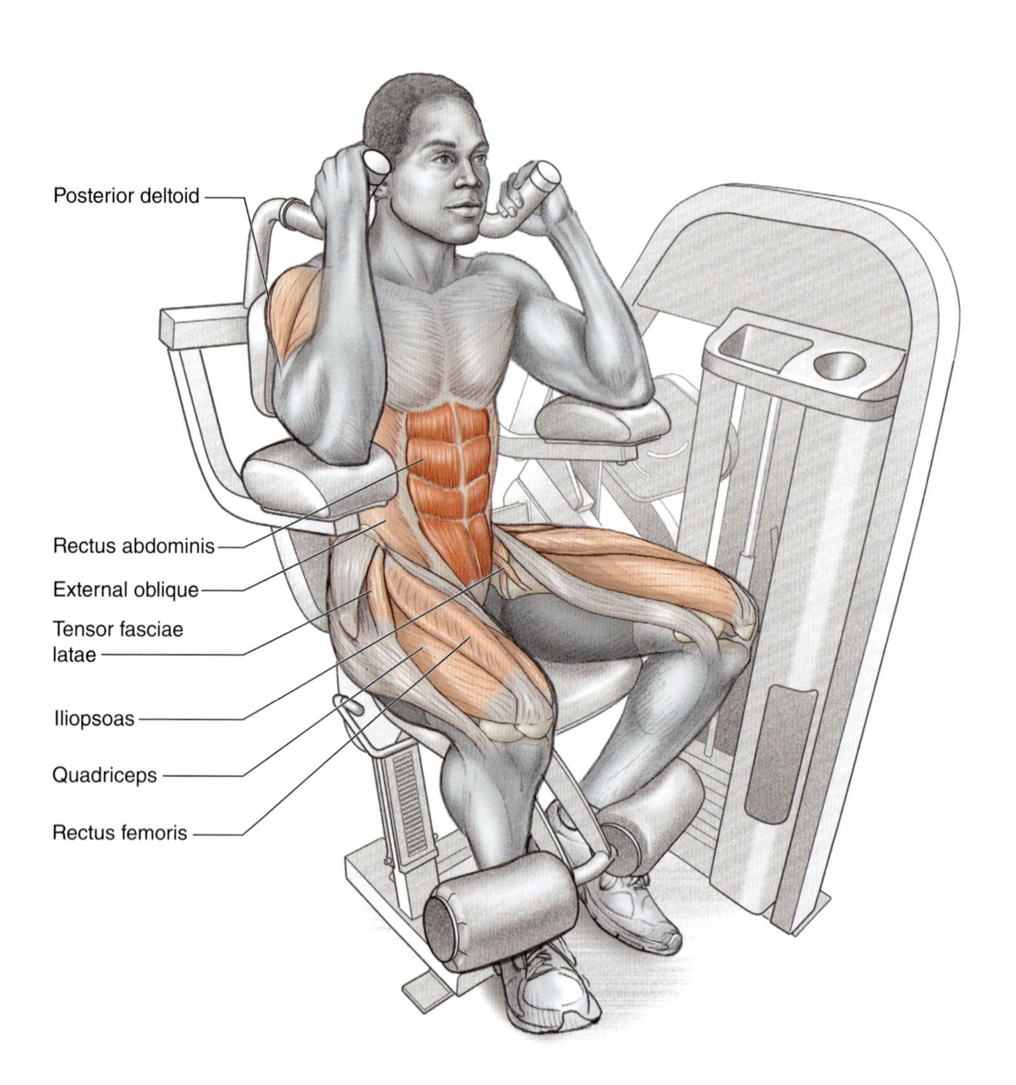

EXECUTION

1. Find a comfortable position in the crunch machine by placing the machine on the setting that most stretches your abdomen, while still allowing you to have enough of a purchase on the machine that you can generate a lot of force.
2. Bring your elbows down to your knees as you crunch your abdomen down and forward, making an arcing motion with your torso.
3. Once you cannot go down any lower (either because you've touched your elbows to your thighs or because the machine's range of motion has run out), hold for a second. Then arc back slowly and with control the same way you came down, getting a big stretch at the top before beginning another rep.

MUSCLES INVOLVED

Primary: Rectus abdominis

Secondary: Iliopsoas, rectus femoris, sartorius, tensor fasciae latae, latissimus dorsi, posterior deltoid, external oblique, internal oblique, transversus abdominis, quadriceps (rectus femoris, vastus lateralis, vastus medialis, vastus intermedius), erector spinae (iliocostalis, longissimus, spinalis)

EXERCISE NOTES

Focus on a powerful concentric phase, accelerating your elbows into your knees and slowing down for a gentle pause just before you hit. Machine crunches are very stable and thus conducive to the production of high forces. After pausing and squeezing your abs at the bottom for one or two seconds, stay engaged with them throughout the eccentric phase, to maximize the effect of the exercise. When coming down, think of curling your body as far forward as possible, as if to touch the backs of your elbows to the fronts of your knees. This arcing motion will help you activate your abs and not just your hip flexors. If you find that you're too fatigued after some reps to create this arc (and you're now just bending at the hips), that means you've hit technical failure and should stop the set to rest for the next one.

TIMING CONSIDERATIONS

You'll need a machine for this, so you may need to consider other exercises if you have limited equipment or are working out in a busy gym. Because this exercise is heavily loadable, it's a good choice for very strong athletes.

SAFETY

When at the top stretch of the movement, keep your abs tense and your spine slightly flexed. Don't relax at the top; just dig in for the next rep!

VARIATIONS

Play with the seat height on the machine so that you can get as much range of motion as possible but still feel strong and connected to your abs throughout. Some crunch machines will feel off until and unless you adjust the seat to your ideal height.

CABLE CRUNCH

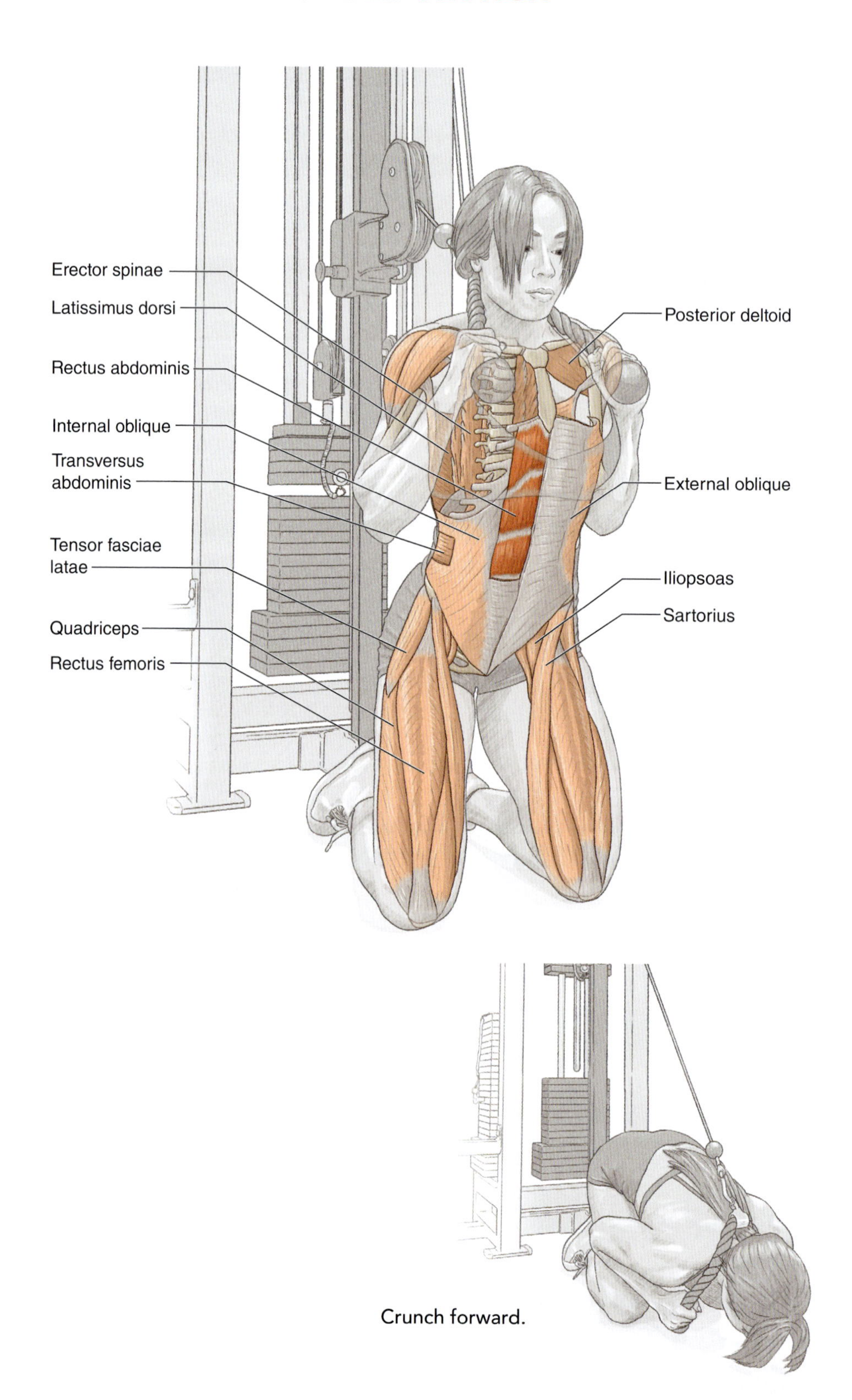

Crunch forward.

EXECUTION

1. Hold a cable attachment of your choice (such as a straight or EZ bar or the rope) just above your head while kneeling on the ground about three to four feet (one meter) in front of the cable stack. There should be enough height to the cable that you can pull all the slack out when sitting on your knees with your thighs and upper body oriented completely vertically.
2. Arcing down and forward, aim to touch your elbows as close as possible to the ground just in front of your knees. Execute this motion by actively curling your abdomen and flexing your hips; try to minimize arm movement.
3. Once you touch the ground with your elbows, pause for a second. Then slowly return for a big stretch before starting another repetition.

MUSCLES INVOLVED

Primary: Rectus abdominis

Secondary: Iliopsoas, rectus femoris, sartorius, tensor fasciae latae, latissimus dorsi, posterior deltoid, external oblique, internal oblique, transversus abdominis, quadriceps (rectus femoris, vastus lateralis, vastus medialis, vastus intermedius), erector spinae (iliocostalis, longissimus, spinalis)

EXERCISE NOTES

Tension under stretch is important for growth. If you stay directly under the machine, you lose leverage for tension under stretch at the top, which gives up a lot of hypertrophic stimulus. Instead, walk forward a few feet (about one meter) and then perform the exercise. This will make it much harder but also much more effective. As you crunch, make sure to not just try to crunch down; you should also try to crunch forward, as if painting the biggest arc possible with your elbows on the way down and back up. This will ensure the exercise targets the abdominals as much as it does the hip flexors. It will make the exercise harder to do, but you're trying to get results!

TIMING CONSIDERATIONS

When you get strong or light enough, this will no longer be a viable exercise. At that point your rep strength for this exercise will begin to exceed your body weight, and you'll be unable to touch the ground at all, with your whole body coming off the ground when you pull. If you don't switch to a more challenging exercise (such as the hanging leg raise or modified candlestick), you will be training so far from failure or for such high reps (30 or more) that you won't be getting a very robust stimulus from this exercise any longer.

> *continued*

SAFETY

Don't cut the range of motion at the top of this exercise. Keep your abs just slightly flexed at the top so that you don't go into hyperextension when transitioning from eccentric to concentric phases.

VARIATIONS

Try different grip handles for comfort. Some people like the rope, some like a solid attachment, and some like to explore everything in between.

CHAPTER 6

Legs

A set of powerful and muscular legs are just as coveted as a well-developed chest, back, shoulders, and arms. This chapter focuses on the lower half of the body. The muscles of the legs (figure 6.1) can be broadly divided into five main groups: the glutes, quadriceps, hamstrings, adductors, and calf muscles. Each group contains several individual muscles that perform specific functions.

We begin with the gluteal muscle group, also known as the "glutes." This muscle group is composed of the gluteus maximus, gluteus medius, and gluteus minimus. The gluteus maximus is the largest and most superficial muscle of the group, contributing significantly to the shape and contour of the posterior. Originating from the ilium, sacrum, and coccyx, and inserting onto the iliotibial band and the gluteal tuberosity of the femur, the gluteus maximus is responsible for extension and lateral rotation of the hip. Underneath the gluteus maximus lies the gluteus medius, which originates from the outer surface of the ilium and inserts onto the greater trochanter of the femur. The gluteus medius is mainly responsible for abduction and medial rotation of the hip joint and for stabilizing the pelvis during walking. Lastly, the gluteus minimus, the smallest muscle of the three, originates from the outer surface of the ilium and inserts onto the greater trochanter of the femur. The gluteus minimus works in concert with the gluteus medius to abduct and medially rotate the hip and stabilize the pelvis.

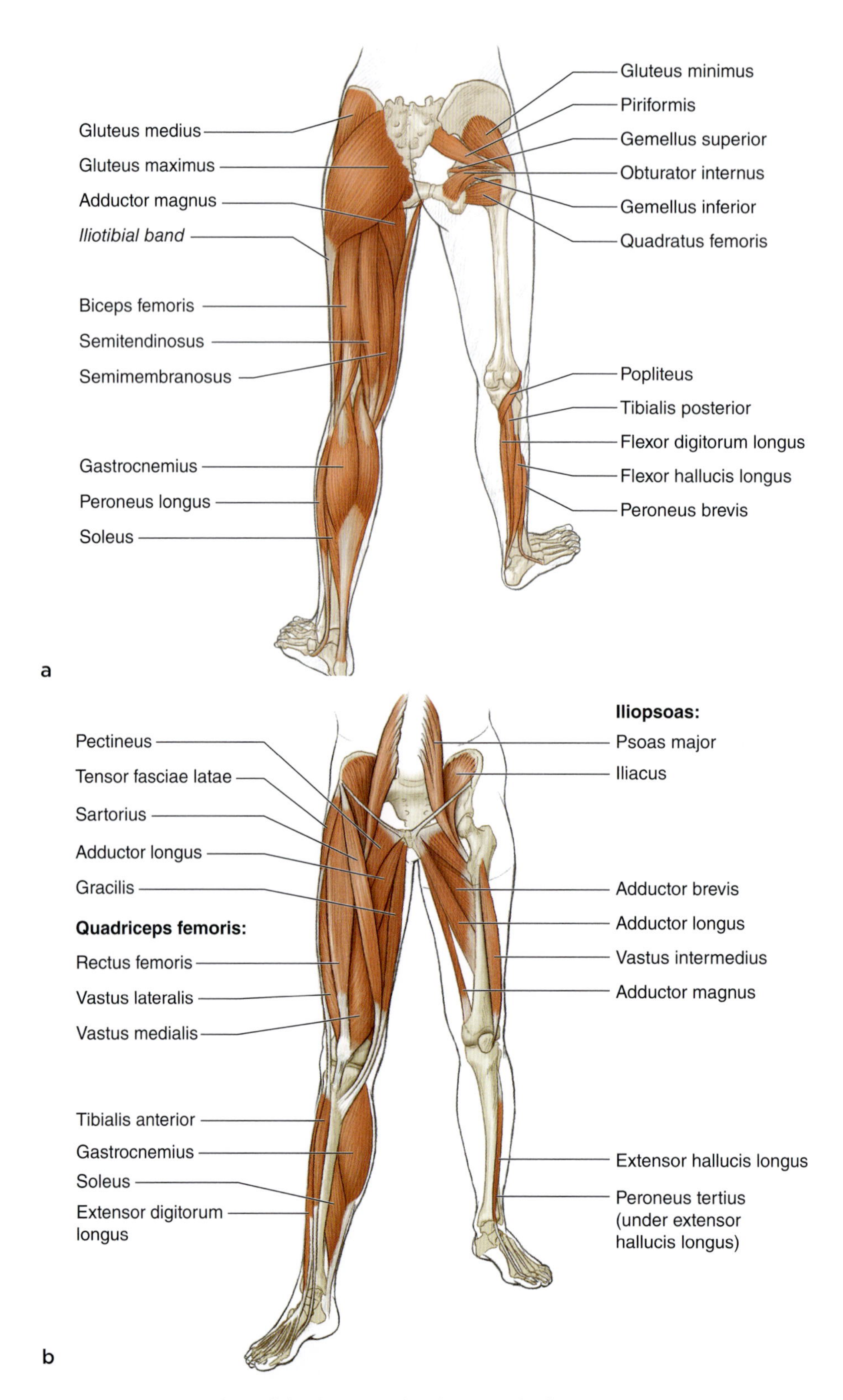

Figure 6.1 Muscles of the legs: (*a*) back view; (*b*) front view.

The quadriceps femoris, also known as the "quads," consist of four distinct muscles on the front of the thigh: rectus femoris, vastus lateralis, vastus medialis, and vastus intermedius. Ultimately, these muscles work synergistically to execute one of the most fundamental movements: knee extension. The rectus femoris originates from the anterior superior iliac spine and the acetabulum of the pelvis and inserts into the patellar tendon; this allows it to uniquely aid in both hip flexion and knee extension because it is the only quad muscle that crosses both the hip and the knee joint. The vastus lateralis originates from the greater trochanter, intertrochanteric line, and linea aspera of the femur and inserts into the patellar tendon. It plays an important role in knee extension and contributes significantly to the overall mass and shape of the thigh. Often recognized for its teardrop shape, the vastus medialis originates from the intertrochanteric line and the linea aspera of the femur and inserts into the patellar tendon. It is responsible for both knee extension and stabilization of the patella. Finally, the vastus intermedius originates from the surface of the femur and inserts into the patellar tendon. It is responsible for knee extension. Because the vastus intermedius lies deep compared to the rectus femoris, it is the least visible of the quadriceps muscles.

The hamstrings, situated at the back of the thigh, are composed of three main muscles: the biceps femoris, semitendinosus, and semimembranosus. These muscles work mainly to provide hip extension and knee extension; they also contribute to overall leg aesthetics and balance. The long head of the biceps femoris originates from the ischial tuberosity in the pelvis, while the short head originates from the linear aspera on the femur. Both heads converge and insert onto the fibular head and lateral side of the tibia. The biceps femoris plays a large role in knee flexion and hip extension; it also contributes to the lateral rotation of the knee. The semitendinosus originates from the ischial tuberosity in the pelvis and inserts onto the medial surface of the tibia. It provides knee extension, hip flexion, and medial rotation of the knee. Finally, the semimembranosus also originates from the ischial tuberosity and inserts onto the medial condyle of the tibia. It also provides knee extension, hip flexion, and medial rotation of the knee, alongside the semitendinosus.

The adductor muscles are a crucial, sometimes overlooked muscle group among bodybuilders. Situated within the inner thigh, the adductors are composed of five muscles: the adductor longus, adductor brevis, adductor magnus, gracilis, and pectineus. These muscles are mainly responsible for movements such as adduction, flexion, and rotation of the thigh. The adductor longus originates at the anterior body of the pubis and inserts on the linea aspera of the femur. It is responsible for adducting the thigh as well as assisting in hip flexion and medial rotation. The adductor brevis originates on the inferior pubic ramus and inserts on the linea aspera of the femur. It is responsible for adduction and medial rotation of the thigh; it is also responsible for hip flexion. The adductor magnus is a large, triangular-shaped muscle that consists of two parts: a pubofemoral (adductor) and ischiocondylar (hamstring) part. The pubofemoral fibers originate from the inferior pubic ramus and insert into the linea aspera of the femur. They are responsible for hip adduction and flexion. The ischiocondylar fibers originate from the ischial tuberosity and insert into the adductor tubercle on the medial condyle of the femur. They are responsible for hip adduction and extension, similar to that of the hamstrings. The gracilis, a thin and flat muscle, is the most superficial muscle on the medial side of the thigh. It originates from the pubic symphysis and inferior pubic ramus and inserts onto the medial shaft of the tibia beneath the medial condyle. The gracilis is involved in hip adduction, flexion, and medial rotation. Lastly, the pectineus is a flat, quadrangular-shaped muscle that originates at the superior pubic ramus and inserts onto the pectineal line of the femur. It is responsible for adduction and flexion of the hip.

Finally, the calf muscles, also known as the "calves," provide lower body strength, stability, and aesthetic appeal. They play a vital role in foot and ankle movement. The calf is composed of two major muscles: the gastrocnemius and the soleus. The gastrocnemius is a superficial, two-headed muscle. Each head originates from the medial and lateral condyles of the femur, and both heads converge into the Achilles tendon. The Achilles tendon then inserts into the posterior surface of the calcaneus and is primarily responsible for plantar flexion of the ankle joint. Underneath the gastrocnemius is the soleus muscle, which originates from the head of the fibula and converges into the Achilles tendon. It aids in plantar flexion of the foot at the ankle joint.

Let's move on to the three main leg joints: the hip, knee, and ankle. The hip joint is a ball-and-socket joint that allows for a wide range of movements between the head of the femur and the acetabulum of the pelvis: flexion, extension, abduction, adduction, internal rotation, external rotation, and circumduction. The knee is a hinge joint located between the femur and tibia; it allows for flexion and extension. The ankle is a hinge joint located between the tibia and fibula in the leg and the talus in the foot; it allows for plantar flexion.

DEFICIT SUMO DEADLIFT

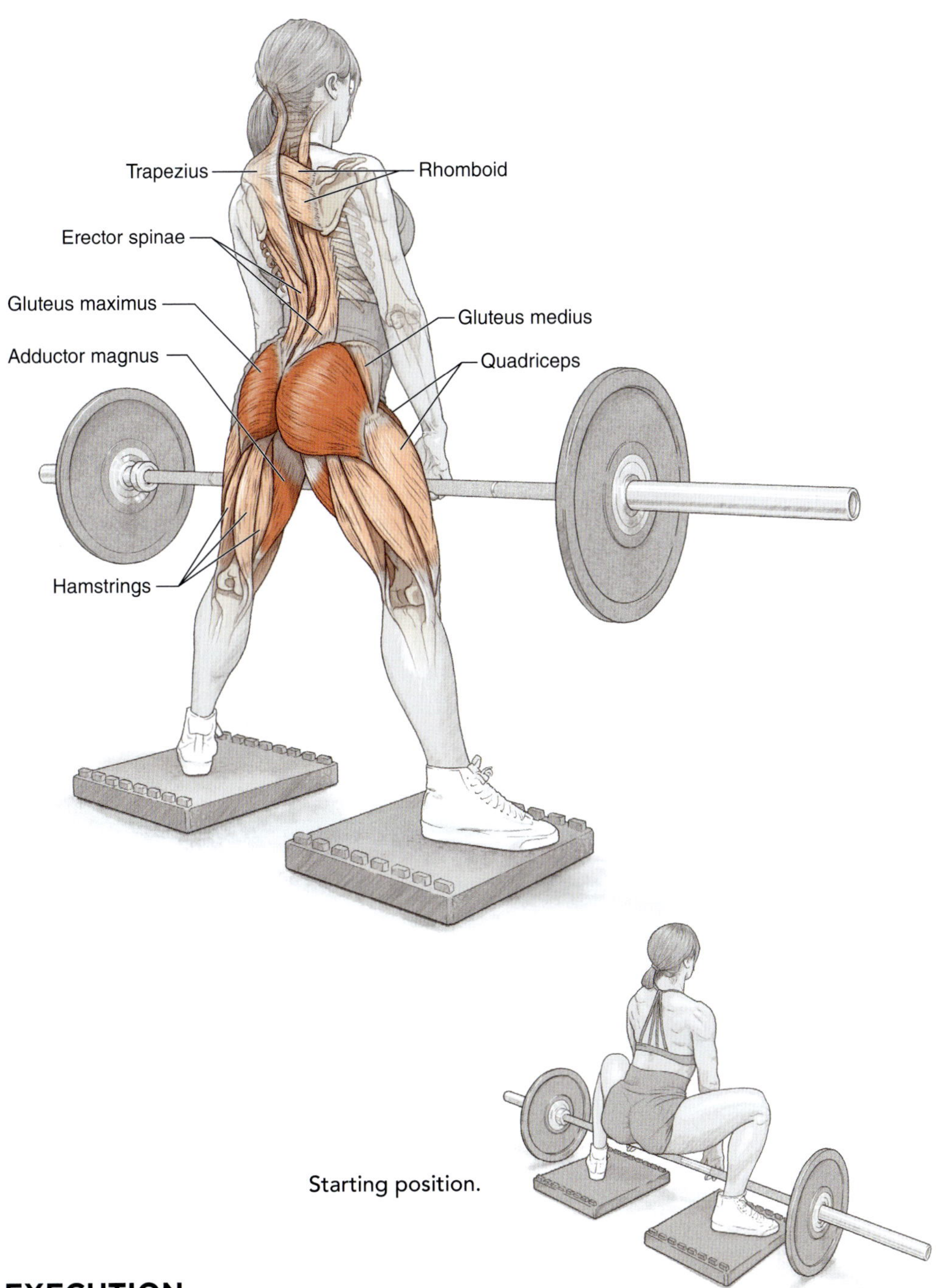

Starting position.

EXECUTION

1. Stand on a two- or three-inch (five- to seven-centimeter) platform, or use 25-pound (10 kg) small plates as the biggest plates on the bar for the deficit effect. Place your feet just outside shoulder width, with your toes pointed out at a 45-degree angle.

> *continued*

2. Keep your chest up and your back flat. Push your knees out as you bend them, and push your hips back as far as they'll go with your lower back still flat.
3. Find the bar with your hands at just inside shoulder width. Grip it, and—with your chest up, your shins flush to the bar, your back flat, and your hips back as far as possible—stand straight up, leading with your chest.
4. Descend slowly in exactly the same manner as the initial setup. Pause on the ground for a second while maintaining tension throughout. Then repeat.

MUSCLES INVOLVED

Primary: Gluteus maximus, adductor magnus, adductor longus, adductor brevis

Secondary: Quadriceps (rectus femoris, vastus lateralis, vastus medialis, vastus intermedius), hamstrings (semitendinosus, semimembranosus, biceps femoris), posterior fibers of the gluteus medius, erector spinae (iliocostalis, longissimus, spinalis), rhomboid, trapezius, teres major, rectus abdominis, external oblique, internal oblique, transversus abdominis

EXERCISE NOTES

On the way up during each rep, keep the knees out, and let the glutes do the work. If your knees start to track in, you may lose a mind–muscle connection with the glutes, and your adductors may end up feeling the movement more. That won't be the case for everyone, but it can help. You may be tempted to rush on the way down, but that slow eccentric is a big deal. Ride it out, and gently pause at the bottom to drive stretch-mediated hypertrophy.

TIMING CONSIDERATIONS

This is a very effective lift for glute hypertrophy; however, it is also axially and systemically fatiguing. You can obviate some of this fatigue by doing hip thrusts or lunges beforehand, so that the glutes become the limiting factors with fewer reps and less weight. Also, because the glutes get a lot of stimulus from most quad training, most uses of this exercise should be saved for a specialization phase targeting the glutes (when the bigger stimulus is really worth the added fatigue).

SAFETY

Keep your lower back tight, and don't bounce the weights off of the ground. To reduce the probability of aggravating your biceps, lift with completely extended (not partially bent) arms.

VARIATIONS

Deeper is better in most cases, but foot stance width and toe point angle are both highly variable and highly individual. Experiment with as many positions as you like before picking one standard to execute for a few months in a row.

SUMO SQUAT

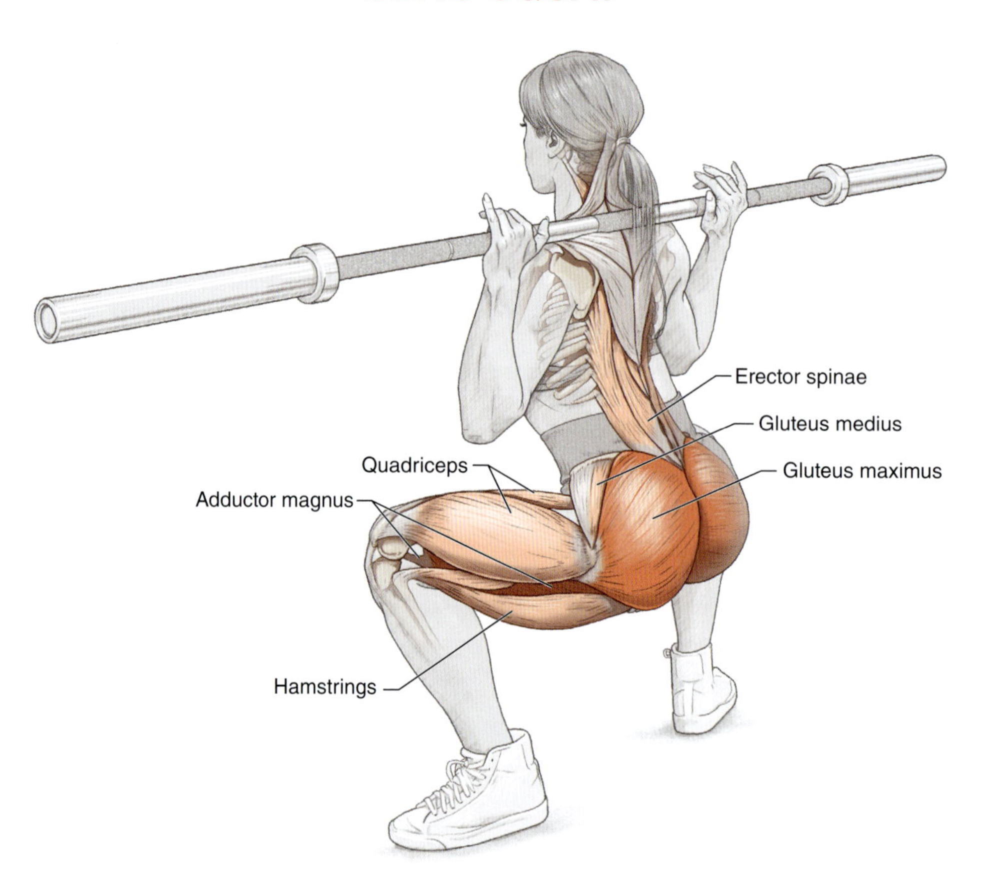

EXECUTION

1. With the bar on your back (right on top of your traps, with your shoulder blades pinched back), walk back out of the rack and place your feet just outside shoulder width. Your toes should be pointed out at a 45-degree angle.
2. Keep your chest up and your back flat. Push your knees out as you bend them, and push your hips back as far as they'll go with your lower back still flat.
3. Descend until your thigh is below parallel to the ground, as far as you can go while maintaining a flat (neutral or lordotic, not kyphotic) back.
4. Hold the bottom position for a second or two. Then stand straight up, leading with your chest.

MUSCLES INVOLVED

Primary: Gluteus maximus, adductor magnus, adductor longus, adductor brevis

Secondary: Quadriceps (rectus femoris, vastus lateralis, vastus medialis, vastus intermedius), hamstrings (semitendinosus, semimembranosus, biceps femoris), posterior fibers of the gluteus medius, erector spinae (iliocostalis, longissimus, spinalis), rectus abdominis, external oblique, internal oblique, transversus abdominis

EXERCISE NOTES

In the regular squat for quads, you squat straight down. In the sumo squat for the glutes, try to squat back as much as down, reaching your glutes as far back behind you while maintaining a tight lower back. Keep your chest up at all times to ensure safety, and pause for a second or two at the bottom. This won't be much fun, but it will grow your glutes.

TIMING CONSIDERATIONS

Although this exercise is highly effective in promoting glute hypertrophy, it can also cause significant overall fatigue. To minimize fatigue, perform hip thrusts or lunges before this exercise. This will ensure that the glutes are the primary focus of the exercise and can grow robustly from fewer repetitions and less weight. Moreover, since the glutes receive stimulation during most quad training exercises, reserve this exercise for a specific period of specialization where the emphasis is on developing the glutes. This approach will make the fatigue and effort worthwhile.

SAFETY

If everything goes well on this lift, the risk of injury is incredibly low. On the other hand, if you get far off track, the risk is notable. Don't rush the descent under any circumstances, and keep your lower back tight and your chest up at all times.

VARIATIONS

You can try a slow eccentric phase (three to five seconds) or experiment with different foot positions and toe angles. As much as it pains us to write this, you can even try this exercise with a light band around your knees, cueing you to use those glutes at all times.

FRONT FOOT ELEVATED SMITH MACHINE LUNGE

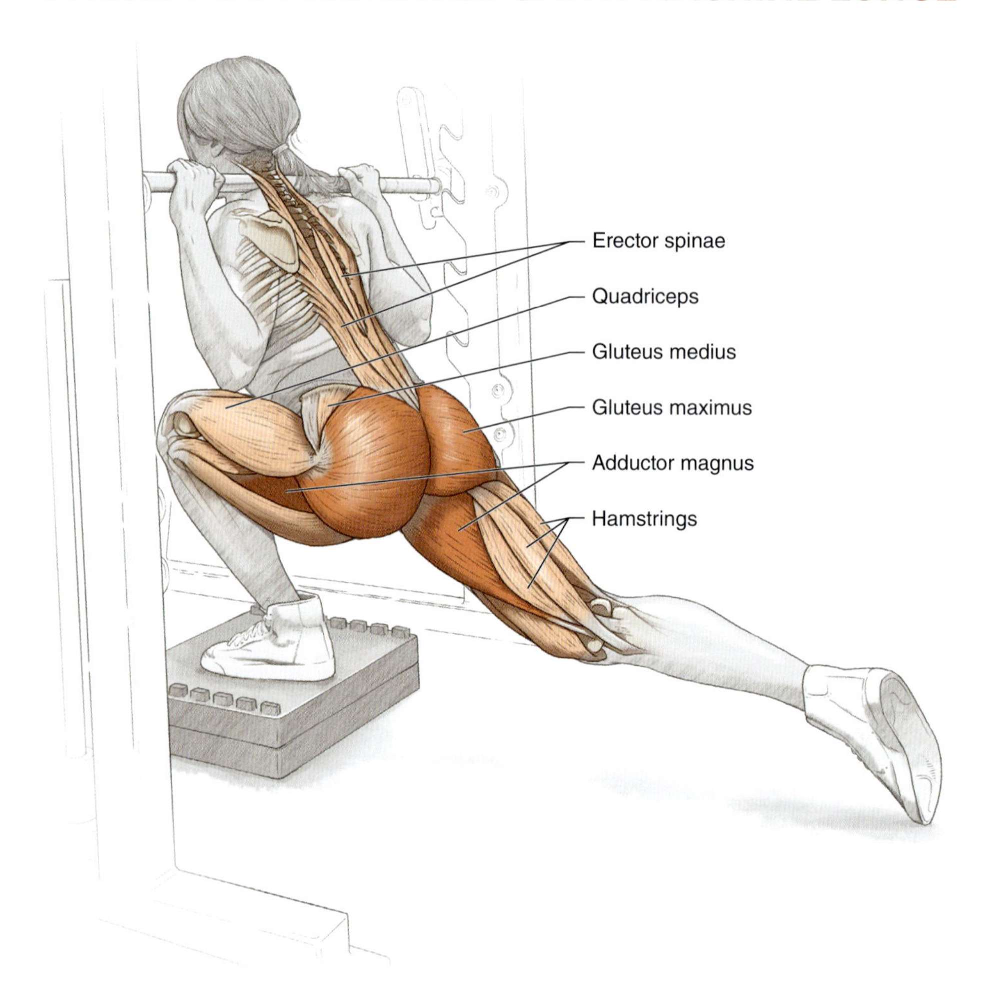

EXECUTION

1. Place a weight plate, mat, or platform on the floor, a few inches (or about five centimeters) in front of the bar path of the Smith machine.
2. Get under the Smith bar, and stand with one foot about one-and-a-half feet (or about half a meter) in front of you on the weight plate, mat, or platform. Place the other foot a similar distance back in the other direction. Grip the bar slightly wider than shoulder-width apart.
3. Keeping your chest up the whole time, slowly descend so that your back knee gently touches the ground behind the platform. Then come back up.
4. After you've completed reps on one leg, rest until you're ready to do another hard set. Then repeat the same technique with your legs switched.

MUSCLES INVOLVED

Primary: Gluteus maximus, adductor magnus, adductor longus, adductor brevis

Secondary: Quadriceps (rectus femoris, vastus lateralis, vastus medialis, vastus intermedius), hamstrings (semitendinosus, semimembranosus, biceps femoris), posterior fibers of the gluteus medius, erector spinae (iliocostalis, longissimus, spinalis), rectus abdominis, external oblique, internal oblique, transversus abdominis

EXERCISE NOTES

The first time you perform this exercise, it might feel a bit weird—but have no fear. If you're patient, you'll find your best front foot position with practice. To maximize the effect on your glutes, push as much as possible with the lead (elevated) leg and as little as possible with the back leg (using it mostly for balance). Lastly, use a slow eccentric phase and pause at the bottom to make this move brutal in the best way possible.

TIMING CONSIDERATIONS

This exercise is so effective that you should perform only one set on each leg the first time you do it. Otherwise the delayed onset muscle soreness (DOMS) can be debilitating and counterproductive. We highly recommend stopping five or six reps short of failure on those first sets and then making it tougher in the following weeks. This exercise is the real deal!

SAFETY

Don't bang your knee on the ground, no matter how much you want the pain of the slow eccentric phase to stop—and definitely don't bounce it off the ground to get more reps. You're not training your knees; you're training your glutes!

VARIATIONS

To really focus on generating force from the front foot, try bringing the back leg off of the ground at the top of the rep; the Smith machine bar will provide stability. Once the next rep begins, you can touch that leg gently to the ground with only your tiptoes and barely put any force on it the whole time.

BARBELL HIP THRUST

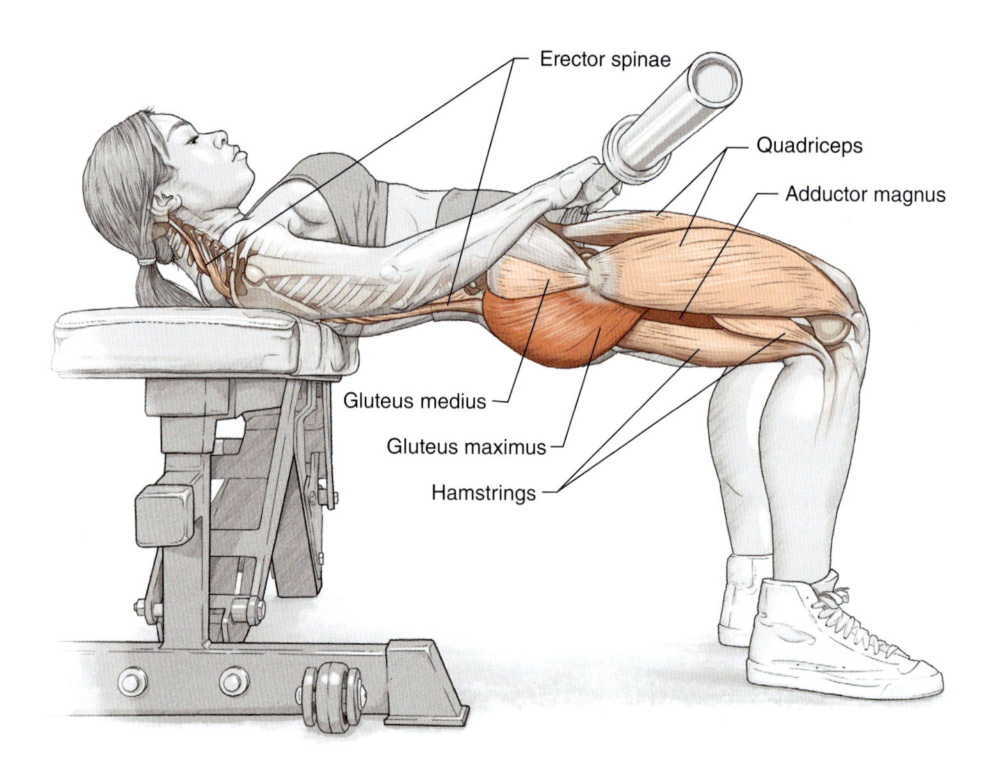

EXECUTION

1. Set up a bench, and roll a barbell with a plate on each side in front of it. Cover the middle of the barbell with a pad or towel.
2. Wiggle under the barbell so that it aligns with your hips. Place your feet shoulder-width apart and your heels and toes firmly on the ground so that your knees form at least a 90-degree angle. Lean your upper back on the bench. Place your hands on the bar slightly wider than shoulder-width apart, depending on your anatomy and comfort.
3. Push your hips up through the bar, lifting the bar until you're at maximum hip extension and cannot go any higher.
4. Slowly lower the load until you no longer feel much tension in your glutes. Then come up again.

MUSCLES INVOLVED

Primary: Gluteus maximus, adductor magnus, adductor longus, adductor brevis

Secondary: Quadriceps (rectus femoris, vastus lateralis, vastus medialis, vastus intermedius), hamstrings (semitendinosus, semimembranosus, biceps femoris), posterior fibers of the gluteus medius, erector spinae (iliocostalis, longissimus, spinalis), rectus abdominis, external oblique, internal oblique, transversus abdominis

EXERCISE NOTES

Because this exercise delivers very low forces at the stretch, peak contraction is the big game here. To maximize it, try to thrust into slight hip hyperextension, pausing there for one to two seconds. Then slowly lower the bar for at least the top third of the movement before lowering a bit faster and then coming back up again. This will reduce how much load you can lift on the exercise but increase its growth-promoting potential. That's a win–win!

TIMING CONSIDERATIONS

Because of the lack of stretch-mediated hypertrophy potential on this exercise, and the high amount of time and sometimes effort it takes to set it up, it's best to reserve the hip thrust for a glute specialization phase. Use it first in the session to pre-exhaust the glutes and create mind–muscle connection; this will ensure you can effectively perform exercises such as lunges and sumo squats after this exercise.

SAFETY

Don't chase loads on this exercise; it's easy to lift a lot of weight with mediocre technique and then gain little benefit. Instead, focus on the technique, reps, and modalities on this exercise (such as myo-reps) that maximize pumps and the mind–muscle connection to the glutes.

VARIATIONS

You can try longer pauses on this exercise, different foot widths or distances from the bar, and different machines that replicate the motion. Some well-built glute machines offer a better version of this exercise than free weights, especially machines that load the glutes more at the stretch.

WALKING LUNGE

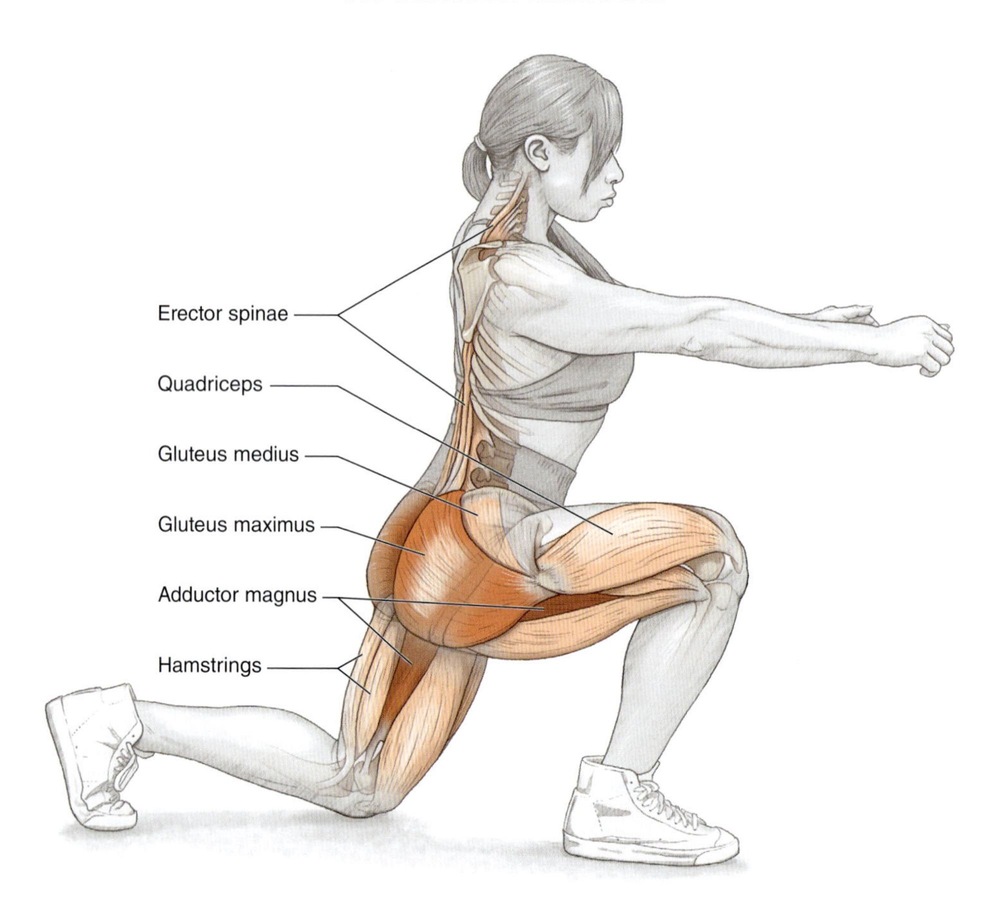

EXECUTION

1. With a barbell on your back or dumbbells in your hands, or using just your body weight, step your right foot as far forward into a lunge as you can while still maintaining balance.
2. Slowly lower your body while keeping your chest up. Gently touch your back knee to the ground.
3. Once there, come up and forward so that your back leg steps beside your leading leg. Take the next step with the left leg.

MUSCLES INVOLVED

Primary: Gluteus maximus, adductor magnus, adductor longus, adductor brevis

Secondary: Quadriceps (rectus femoris, vastus lateralis, vastus medialis, vastus intermedius), hamstrings (semitendinosus, semimembranosus, biceps femoris), posterior fibers of the gluteus medius, erector spinae (iliocostalis, longissimus, spinalis), rectus abdominis, external oblique, internal oblique, transversus abdominis

EXERCISE NOTES

Take the longest steps you can while remaining stable; this will increase the degree of both glute and quad flexion in the lead leg and maximize the growth stimulus. If you have good stability, focus on pushing with mostly your lead leg; that is the leg that will stimulate the glute gains via its loaded stretch. A slow eccentric phase and gentle reversal (after the knee touches the ground) from the bottom will pay dividends, even though they will be super tough to pull off during the set.

TIMING CONSIDERATIONS

This exercise is easy to execute because you need zero equipment to do it right. In fact, you can superset this exercise after any other glute or quad exercise ends for some real fun.

SAFETY

Don't bang your knee on the ground, no matter how much you want the pain of the slow eccentric phase to stop—and definitely don't bounce it off the ground to get more reps. You're not training your knees; you're training your glutes!

VARIATIONS

It is fine to take a whole step, bringing the lagging leg over for another leading step without touching the ground by your other foot. It is also fine to take a staggered step, where you pause with your feet together before stepping forward. As you tire within a set or begin to use more load over months and years of muscle growth, staggering your steps will become the best way to execute this exercise.

Resting a few seconds between steps when you're near failure can lengthen a set by giving you a few extra reps at the end when you're close to failure. This is an amazing and time-saving way to get a bit more growth.

STIFF-LEGGED DEADLIFT

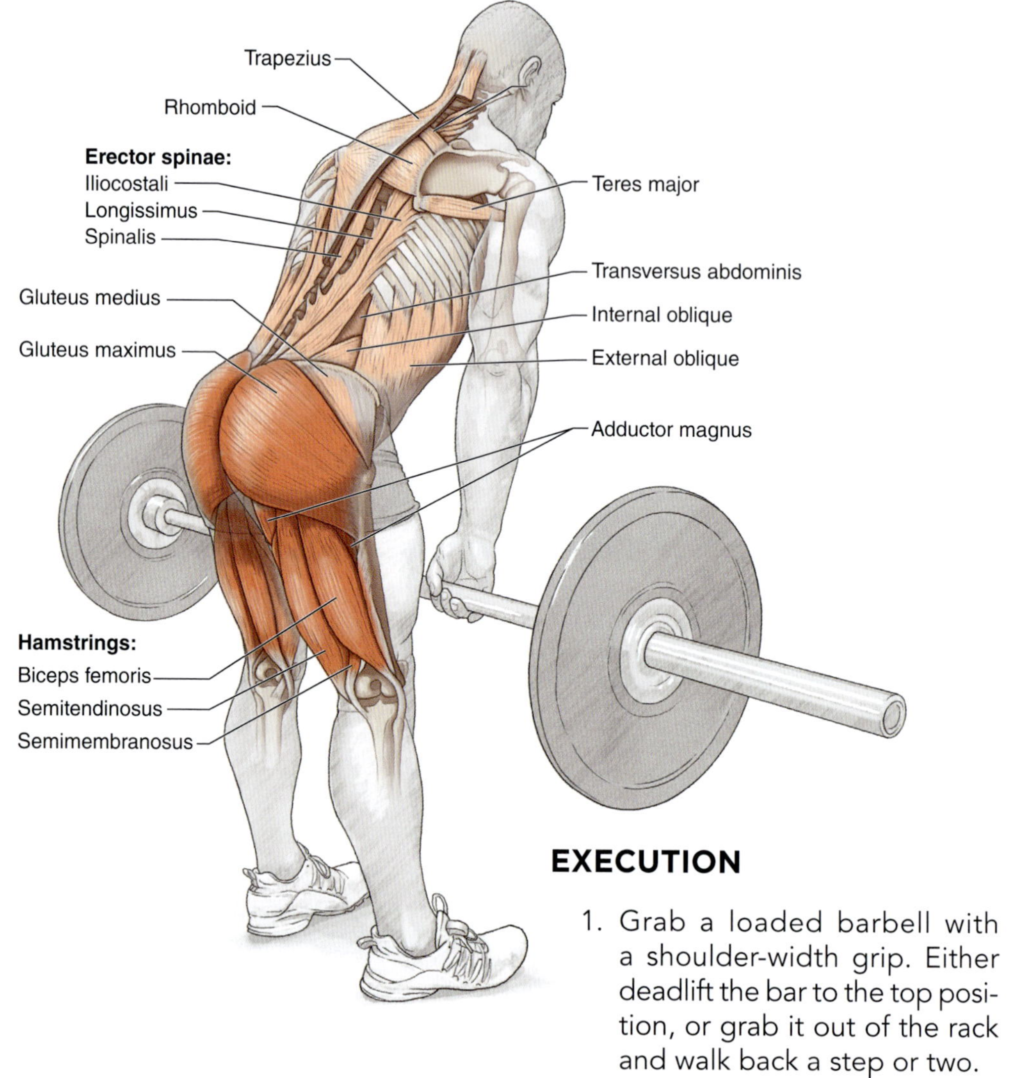

EXECUTION

1. Grab a loaded barbell with a shoulder-width grip. Either deadlift the bar to the top position, or grab it out of the rack and walk back a step or two.
2. With your feet just inside shoulder-width apart and your toes pointed slightly out, begin the movement by moving your hips back while keeping your chest high and your lower back in lordosis (anteriorly rotating your pelvis).
3. Break by about 10 degrees in your knees and keep that bend in your knees through the rest of the movement. Continue back with your hips and your chest up, and slowly ease into greater degrees of hip flexion.
4. Once you go low enough to experience a lot of acute, stretch-mediated hamstring muscle pain—or once you can no longer descend without either rounding your lower back or bending your knees more—come back up by at first gently accelerating up and then moving athletically back to the starting position to begin another rep.

MUSCLES INVOLVED

Primary: Hamstrings (semitendinosus, semimembranosus, biceps femoris), gluteus maximus, adductor magnus, adductor longus, adductor brevis

Secondary: Posterior fibers of the gluteus medius, erector spinae (iliocostalis, longissimus, spinalis), rhomboid, trapezius, teres major, rectus abdominis, external oblique, internal oblique, transversus abdominis

EXERCISE NOTES

When you tilt your pelvis toward the anterior, the hamstrings will be prestretched before any hip flexion occurs. This will ensure that any further hip flexion stretches the hamstrings faster, minimizing the range of motion and erector spinae and glute activity required to stimulate the hamstrings in this exercise. One of the easiest cues for an anterior pelvic tilt is to arch the lower back and keep the chest up and out. For most people, this cue will let them place their whole back into slight lordosis and allow them to enter a very natural anterior pelvic tilt. To continue to flex the hips while maintaining anterior tilt, you can think of trying to show your glutes to a camera positioned just at the corner of the wall and the ceiling directly behind you. This should result in direct knee flexion that attempts to push your knees out of extension, reducing the stretch on the hamstrings. The "chest up and butt-to-ceiling" cue works best only if reinforced with the "keep the knees back" cue; this urges maintenance of the minimal knee angle and helps ensure that further hip flexion will lengthen the hamstrings under load and drive the most muscle growth. Don't lock your knees; the best position for them is usually just shy of lockout.

TIMING CONSIDERATIONS

The stiff-legged deadlift has a high stimulus-to-fatigue ratio. It also has a high raw stimulus magnitude and high amount of fatigue imposition (especially axial fatigue from the spinal loading). Be sure your back can handle the fatigue. When you prioritize this exercise, consider easing up on regular deadlifts, heavy squats, standing barbell shoulder presses, and high volumes of bent-over rows.

SAFETY

Perform the exercise with your back arched and knees back, descending to a depth that produces a painful stretch sensation. Any lower than this depth will likely result in reduced hamstring tension, rendering the exercise less effective. Moreover, descending lower than this point may compromise safety by promoting rounding of the lower back, particularly when lifting heavy loads.

VARIATIONS

You can do stiff-legged deadlifts with many different grip positions as well as with dumbbells, the Smith machine, and loads of other implements. Enjoy each variant for a few months, and then consider switching to another.

LOW-BAR GOOD MORNING

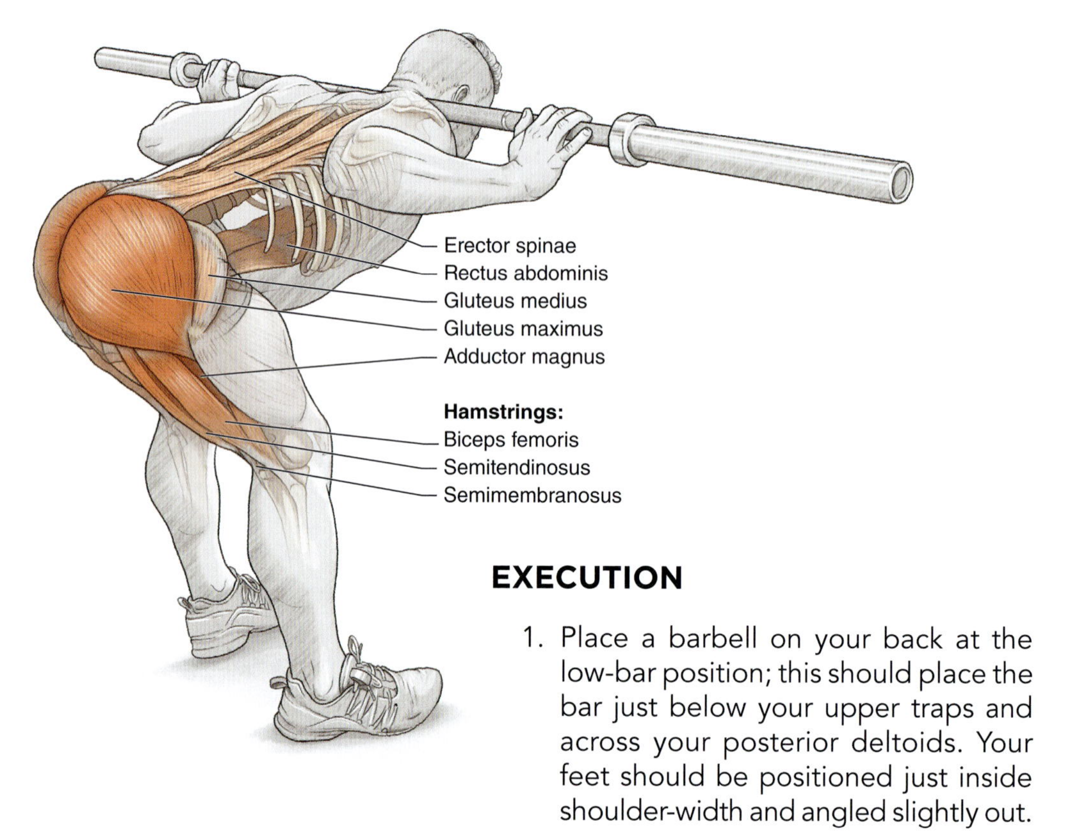

EXECUTION

1. Place a barbell on your back at the low-bar position; this should place the bar just below your upper traps and across your posterior deltoids. Your feet should be positioned just inside shoulder-width and angled slightly out.
2. Initiate the movement by pushing your hips back while keeping your chest high. Hold an anterior pelvic tilt to maintain the natural curvature of your lower back. Bend your knees by about 10 degrees, and maintain that position throughout the entire movement.
3. Keep moving your hips back while keeping your chest up, gradually increasing the hip flexion angle.
4. Once you reach the point where you feel an intense, acute stretch in your hamstrings (or you can no longer descend without rounding your back or bending your knees further), come back up by gently accelerating and then returning to the starting position. Repeat for another repetition.

MUSCLES INVOLVED

Primary: Hamstrings (semitendinosus, semimembranosus, biceps femoris), gluteus maximus, adductor magnus, adductor longus, adductor brevis

Secondary: Posterior fibers of the gluteus medius, erector spinae (iliocostalis, longissimus, spinalis), rectus abdominis, external oblique, internal oblique, transversus abdominis

EXERCISE NOTES

When you tilt your pelvis forward, the hamstrings will be stretched before any hip flexion occurs. This will result in a quicker and more effective stretch on the hamstrings, reducing the range of motion and the amount of erector spinae and glute activity required to stimulate the hamstrings during the exercise. A helpful cue to achieve an anterior pelvic tilt is to arch your lower back while keeping your chest up and out. This will encourage a slight lordosis in your back and a natural anterior pelvic tilt. To maintain the anterior tilt while continuing to flex your hips, imagine trying to present your glutes to a camera located at the corner where the wall meets the ceiling behind you. These cues will promote hip flexion and anterior tilt as you lower but will likely also develop tension in the hamstrings; the result is a direct force that pushes your knees out of extension and reduces the stretch on the hamstrings. Combine the "chest up and butt-to-ceiling" cue with the "keep the knees back" cue; this will help you maintain a slight knee bend and help ensure that further hip flexion lengthens the hamstrings under load and promotes the most muscle growth. Don't lock your knees; the best position for them is usually just shy of lockout.

TIMING CONSIDERATIONS

The good morning is a highly effective exercise for stimulating muscle growth, with a high stimulus-to-fatigue ratio. However, it also generates a substantial amount of fatigue due to the high magnitude of loading it imposes on the paraspinal musculature. To avoid overreaching, it is essential to assess your back's ability to handle the resultant fatigue. When prioritizing this exercise, it may be prudent to reduce the intensity or frequency of other exercises that also impose axial fatigue on the spine (such as heavy deadlifts, squats, standing barbell shoulder presses, and high-volume bent-over rows).

SAFETY

Perform the exercise with your back arched and knees back, descending to a depth that produces a painful stretch sensation. Any lower than this depth will likely result in reduced hamstring tension, rendering the exercise less effective. Moreover, descending lower than this point may compromise safety by promoting rounding of the lower back, particularly when lifting heavy loads. It can help to place the racks at the bottom position; by touching them gently you'll be able to tell when the bottom-end range of motion is achieved and know to come back up; this will standardize your reps for more effective tracking and future overload presentation and fatigue detection.

VARIATIONS

Though low-bar good mornings are highly effective, high-bar and safety bar good mornings offer excellent alternatives if you want to try them and if you have the equipment.

SEATED LEG CURL

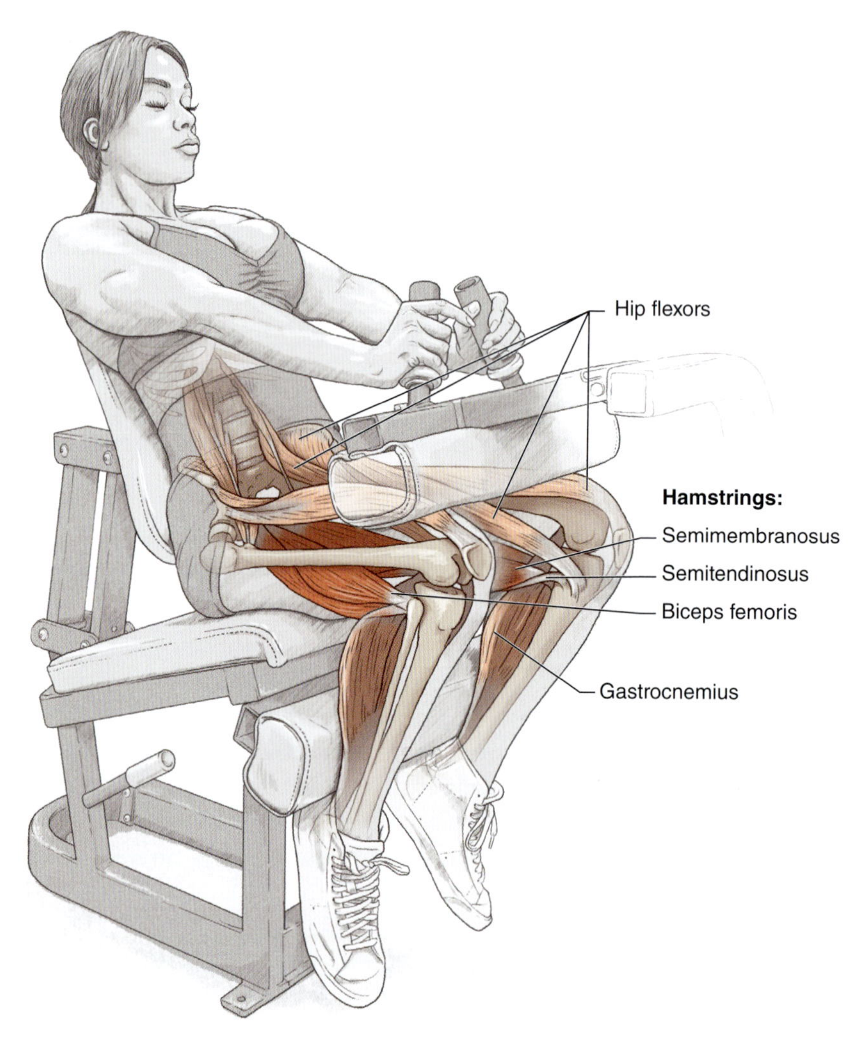

EXECUTION

1. Sit at the leg extension machine and adjust the pads so the foot pads are touching at the very distal ends of your lower legs.
2. Set the top pad as tight to your quads as possible so, at the start of the movement, your knees are extended as much as possible. Begin to accelerate the weight down in an arcing motion.
3. Move the load powerfully, with no emphasis on capping your resultant concentric velocity, until you hear an audible clank of the machine coming together at its terminal range of motion, or until your hamstrings and calves are flush against one another and you cannot generate any more range of motion.
4. Return to the starting position very slowly and with a lot of eccentric control, easing into the starting position gently before repeating for the next rep.

MUSCLES INVOLVED

Primary: Hamstrings (semitendinosus, semimembranosus, biceps femoris)

Secondary: Gastrocnemius, gluteus maximus, gluteus medius, gluteus minimus, iliopsoas, rectus femoris, sartorius, tensor fasciae latae

EXERCISE NOTES

Because the hamstrings are well positioned for eccentric loading, and because such loading is generally effective for promoting muscle growth, it is critical to go slowly during this phase so you can get the most out of this exercise. When you reach the end of the eccentric portion, either hold the position for a second or two or gently accelerate out of it. Abrupt changes in direction are not conducive to promoting muscle hypertrophy and may increase the risk of injury. At the end of the concentric portion, make sure to touch the pad to the end of the machine's range of motion or place your calves flush to your hamstrings; don't pause there in most cases, because pausing will waste energy in what might be the least growth-promoting position of the entire movement. If you're going to pause on this exercise, pause at the position of biggest stretch.

TIMING CONSIDERATIONS

This exercise generates low axial and systemic fatigue, so you can do it at the beginning of a leg training session without compromising much on the performance of later quad and glute exercises. However, if you do this exercise after a lot of quad or glute work, you may find your performance highly degraded because quad and glute training generally creates high systemic fatigue. When including leg curls in their agenda, most people arrange their leg training with the leg curls first to get the most stimulus for the legs as a whole.

SAFETY

Although this exercise is considered safe, it is recommended to slow down the eccentric phase, particularly when approaching the bottom position, to minimize any risk of injury. Avoid rapid pulsing out of the bottom position; instead, gradually accelerate out of the bottom position for optimal safety.

VARIATIONS

You can use different ranges of reps on this exercise to increase variation, at times opting for sets of 10 to 20 reps and sometimes choosing sets of as high as 20 to 30 reps. You can also try some myo-reps, resting in the lockout position for a few seconds a few times in each set to get a few more reps in. Lastly, don't worry too much about your foot positions or which way they point. This overemphasizes detail and makes little difference in effectiveness, yet it also consumes a lot of mental bandwidth (bandwidth that is usually better spent on focusing on the hamstrings instead of the feet).

LYING LEG CURL

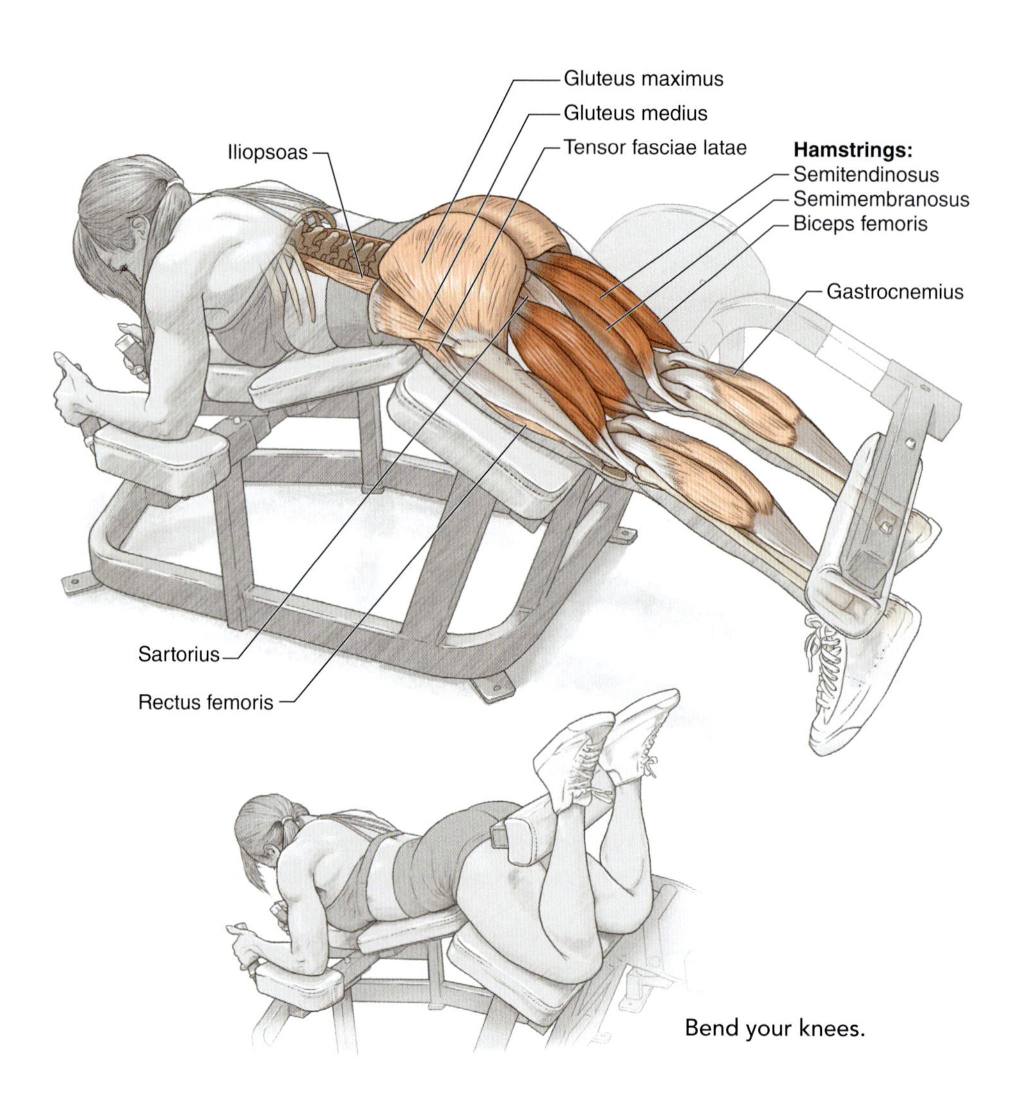

Bend your knees.

EXECUTION

1. Lie face down on the leg curl machine, and adjust the pads so the foot pads are touching at the very distal ends of your lower legs. Adjust the pad angle so you're in a maximum allowable degree of knee extension when you begin.
2. Begin to accelerate the weight in an arcing motion.
3. Move the load powerfully until you hit the pad to your glutes at the top, or until your hamstrings and calves are flush against one another and you cannot generate any more range of motion.
4. Return to the starting position slowly and with eccentric control. Ease into the starting position gently before starting the next rep.

MUSCLES INVOLVED

Primary: Hamstrings (semitendinosus, semimembranosus, biceps femoris)

Secondary: Gastrocnemius, gluteus maximus, gluteus medius, gluteus minimus, iliopsoas, rectus femoris, sartorius, tensor fasciae latae

EXERCISE NOTES

Given the advantageous position of the hamstrings for eccentric loading and the effectiveness of such loading in promoting muscle growth, it is crucial to perform the eccentric phase of this exercise slowly, to maximize its benefits. On reaching the bottom of the range of motion, either hold the position briefly or gradually accelerate out of it. Abrupt changes in direction are not conducive to promoting muscle hypertrophy and may increase the risk of injury. When ascending, aim to touch the pad to your glutes or, if that's not possible due to body proportions or machine design, bring your calves in close proximity to your hamstrings. Pausing at this position is not recommended; it is likely the least growth-promoting part of the movement, and pausing wastes energy. If a pause is warranted, it should be at the point of maximal stretch during the exercise.

TIMING CONSIDERATIONS

This exercise generates low axial and systemic fatigue, so you can do it at the beginning of a leg training session without compromising much on the performance of later quad and glute exercises. However, if you do this exercise after a lot of quad or glute work, you may find your performance highly degraded because quad and glute training generally creates high systemic fatigue. When including leg curls in their agenda, most people arrange their leg training with the leg curls first to get the most stimulus for the legs as a whole.

SAFETY

This is a safe exercise. However, you should still slow down the eccentric phase, especially close to the bottom position, to ensure you're being as safe as possible. In addition, do not pulse out of the bottom position, but instead gently accelerate out of it.

VARIATIONS

If you can't get maximum knee extension at the bottom because of the design of the machine, consider putting a pad or folded yoga mat between your thighs and the machine. This will increase knee extension and maximize stretch-mediated hypertrophy.

GLUTE-HAM RAISE

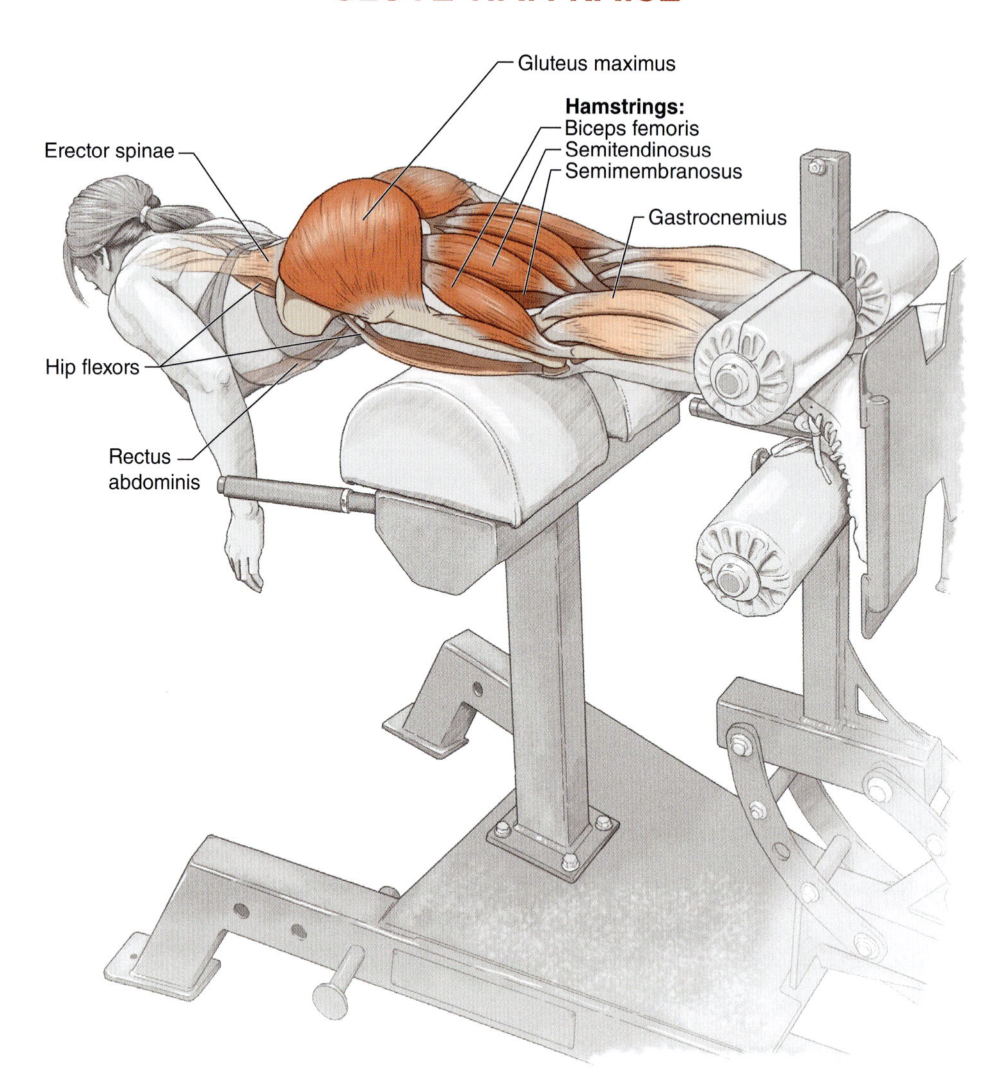

EXECUTION

1. Place your feet into the back supports of a glute-ham machine, keeping your calves flush to the top pad. Place your knees on the main rotation pad, ideally high enough so your hip joints can clear the front of the pad on the way down.
2. Keeping your chest up and back flat, reach your body forward and arc down, extending at your knees while flexing at your hips.
3. Slowly descend as far as the machine allows you, maintaining a neutral spine. Come back up, accelerating powerfully until your knees are flexed and your back and upper legs (as one rigid segment) are perpendicular to the ground.

MUSCLES INVOLVED

Primary: Hamstrings (semitendinosus, semimembranosus, biceps femoris), gluteus maximus, gluteus medius, gluteus minimus

Secondary: Gastrocnemius, erector spinae (iliocostalis, longissimus, spinalis), iliopsoas, rectus femoris, sartorius, tensor fasciae latae, rectus abdominis, external oblique, internal oblique, transversus abdominis

EXERCISE NOTES

To make the exercise as challenging as you can, aim to stretch the hamstrings as much as possible and present the worst possible lever arm for them. Try to reach forward with your chest on the descent, lengthening the hamstrings maximally as you move. When you come back up, repeat that motion in reverse by arcing forward at first as you come up, instead of pulling back. Pulling back is fine, but it reduces the very hamstring forces and stretch you're trying to maximize with this exercise. As soon as you descend into your next rep, engage strict eccentric control right away with slow movement on the descent, to drive the most stimulus and reduce the risk of injury.

TIMING CONSIDERATIONS

Executing this exercise safely and effectively requires a high level of strength relative to body weight. As such, this exercise is best suited for advanced lifters, especially those with shorter torsos in proportion to their upper legs or those who have a lighter body weight. If you wish to incorporate this exercise into your routine, it may be beneficial to first focus on building strength through loaded 45-degree back raises to prepare your body.

SAFETY

You can go fast on the way up by accelerating from a gentle start at the bottom, but you should avoid going fast on the way down to both minimize injury risk and maximize eccentric loading. Because this is a tough exercise, it will be very tempting to rush the eccentric phase (especially as you tire toward the end of a set); try to resist as much as you can!

VARIATIONS

Most setups for glute-ham raises provide a lot of options for placement of your knees and shins. When choosing an option, avoid a range of motion so short that your hips run into the pad at the bottom, resulting in spinal flexion instead of hip flexion; this removes tension from the targeted hamstrings.

NORDIC CURL

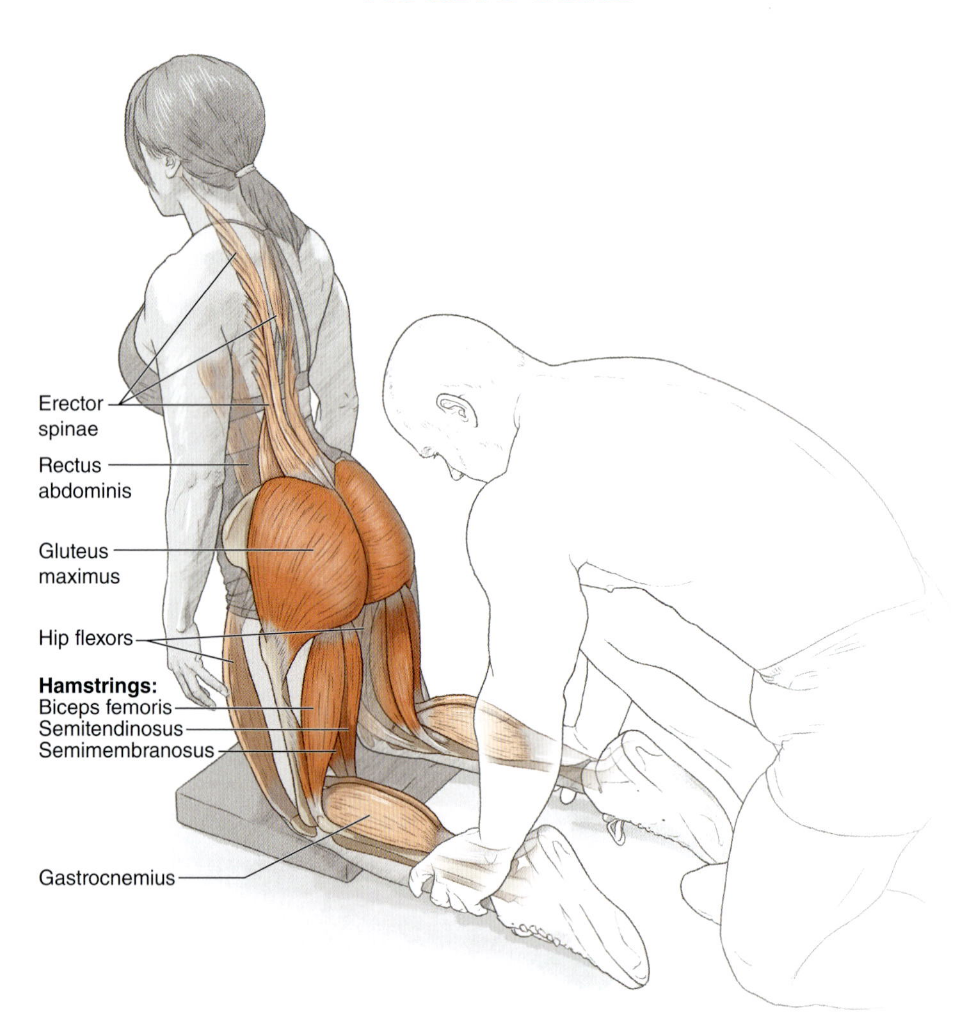

EXECUTION

1. Place your feet tightly under the support pads of Nordic curl machine (or have a partner hold your ankles), and position your knees comfortably on the bottom pad.
2. With your chest held high and your upper body and upper legs forming a straight, rigid segment, slowly descend until your chest touches the pad.
3. Hold the lowered position under tension for a second. Then gently accelerate up until your torso and upper legs are perpendicular to the ground to begin another repetition. All of the range of motion for this exercise should come from knee flexion.

MUSCLES INVOLVED

Primary: Hamstrings (semitendinosus, semimembranosus, biceps femoris), gluteus maximus, gluteus medius, gluteus minimus

Secondary: Gastrocnemius, erector spinae (iliocostalis, longissimus, spinalis), iliopsoas, rectus femoris, sartorius, tensor fasciae latae, rectus abdominis, external oblique, internal oblique, transversus abdominis

EXERCISE NOTES

This exercise is designed to impart eccentric loading, so never rush on the way down. Descend slowly and with a high degree of control. Pause at the bottom of every rep; maintain maximum hamstring tension during that pause so you can net as much stretch-mediated hypertrophy benefit as you can. Especially on the way back up, maintain completely rigid hips; resist the temptation to shift your hips back to reduce the moment arm and hamstring tension. Although such a move is natural and will make the exercise easier, your aim is to make the exercise difficult for the hamstrings so you can promote an unnatural amount of growth in them.

TIMING CONSIDERATIONS

Proper and safe execution of this exercise requires a high strength-to-weight ratio. This exercise is ideally suited for advanced lifters, especially those with shorter torsos in relation to their upper legs and individuals with lighter body weight. It may help to get really strong on loaded 45-degree back raises first before giving this exercise a shot.

SAFETY

The forces imparted on the hamstrings increase linearly with the degree of extension, so the risk for injury on this exercise is low—but definitely not nominal. Do not rush the eccentric phase, especially close to the bottom of the exercise, as that significantly increases the chance of injury.

VARIATIONS

If you're not strong enough to use this exercise traditionally yet, you can place your hands out in front of you as if in a close-grip push-up position. As you descend, your hands will contact the pad and slow you down so that you can get the level of hamstring tension you desire. Stronger individuals can also use this technique to extend their traditional working sets to beyond muscular failure by cranking out a couple of push-assisted reps (especially if they do the eccentric phase with no help and use their arms to push-assist the concentric phase).

REVERSE-HACK GOOD MORNING

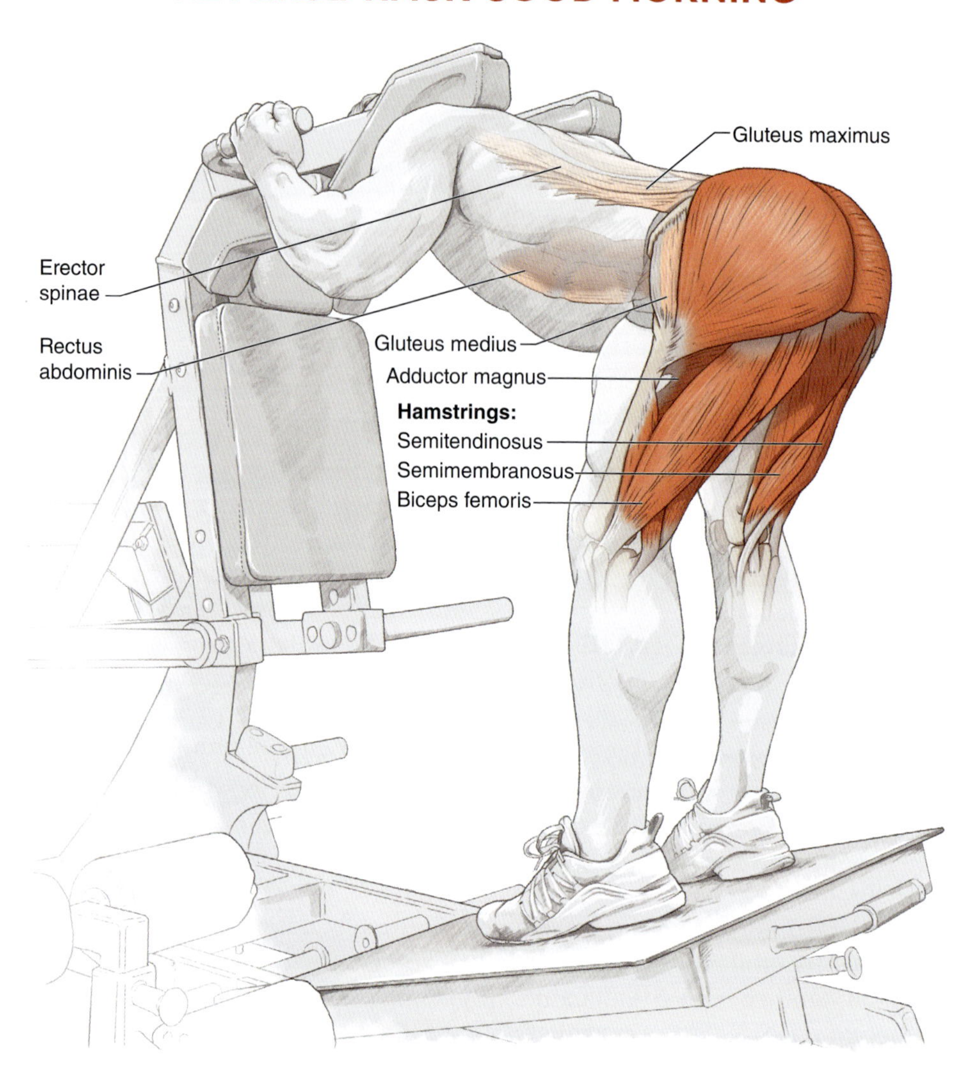

EXECUTION

1. Position yourself into a reverse hack machine (otherwise known as a V-squat machine), and unrack the weight. Place your feet just inside shoulder-width with your toes slightly pointed out.
2. Start the movement by pushing your hips back while maintaining a high chest and anterior pelvic tilt; keep your lower back in lordosis. Bend your knees slightly, maintaining that position throughout the movement. Continue moving your hips back while keeping your chest up and gradually increasing hip flexion.
3. Descend until you feel significant, acute, stretch-mediated pain in your hamstring muscles, or until you can no longer lower yourself without rounding your lower back or bending your knees further.
4. To return to the starting position, gently accelerate upward. Then move powerfully to complete another repetition.

MUSCLES INVOLVED

Primary: Hamstrings (semitendinosus, semimembranosus, biceps femoris), gluteus maximus, adductor magnus (hamstring portion)

Secondary: Posterior fibers of the gluteus medius, erector spinae (iliocostalis, longissimus, spinalis), rectus abdominis, external oblique, internal oblique, transversus abdominis

EXERCISE NOTES

An anterior pelvic tilt performed during hip flexion places the hamstrings into a pre-stretched state; this reduces the required range of motion and the associated spinal erector and glute activation. A great technique to achieve anterior pelvic tilt involves lumbar extension and thoracic extension, encouraging slight lordosis of the lumbar spine. While descending during the exercise, promote hip flexion and anterior tilt by imagining you are presenting your glutes to a camera situated at the junction of the wall and the ceiling behind you. This maneuver will generate high tension in the hamstrings, producing a forceful impulse that extends the knees and, consequently, compromises hamstring stretch. Therefore, by reinforcing the "chest up/butt-to-ceiling" cue with the "keep the knees back" cue, and by maintaining a minimal degree of knee flexion during hip flexion, you can enhance the longitudinal stretch of the hamstrings and optimize muscle growth signaling. You should not try to completely achieve knee joint extension; instead, maintain the knee joint just proximate to full extension.

TIMING CONSIDERATIONS

It's easier to set this exercise up on a machine if one is free (rather than doing a barbell good morning, which is less axially and systemically fatiguing). If you have access to a V-squat machine, this is definitely an exercise to try.

SAFETY

Make sure the support is in when you rack the weight, because failing to place the support under the load can turn a rack attempt into a giant disaster. This is something that will require more attention than usual, since you'll be racking at the end of a set—precisely when you're the most fatigued and least attentive!

VARIATIONS

In most cases, using bands for hypertrophy training is suboptimal because it reverses the force curve best suited for generating muscle growth. Instead of emphasizing the stretch and easing up on the peak contraction, bands usually emphasize the peak contraction and deemphasize the stretch. However, on this machine, the top end (peak contraction) is so light and the force curve so steeply biased to the stretch that using some bands on the machine can make the force curve more even and still considerably stretch-biased; this improves the muscle-growing characteristics of this setup.

HIGH BAR SQUAT

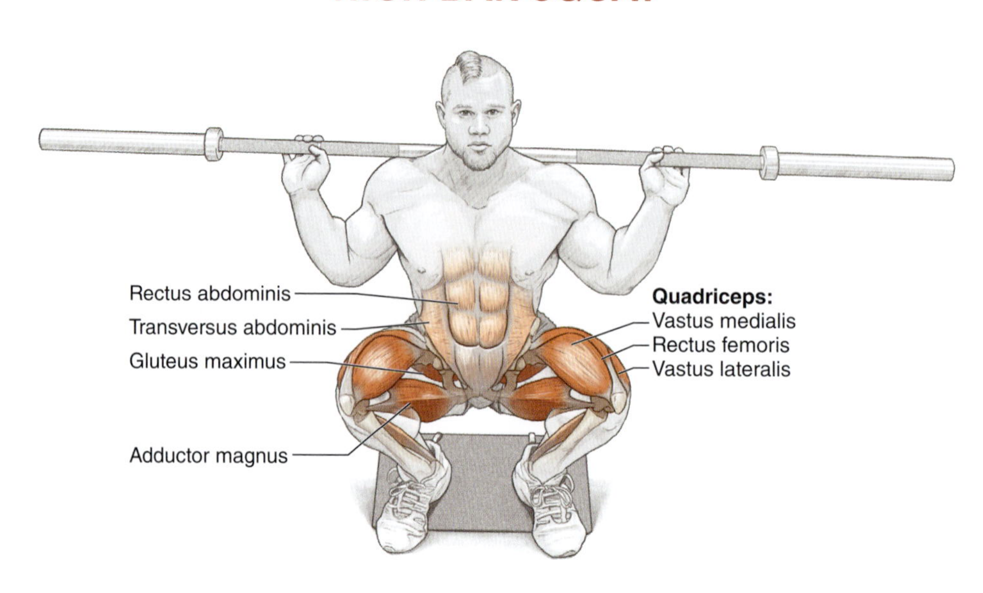

EXECUTION

1. With the bar on your back (right on top of your traps with your shoulder blades pinched back), walk back out of the rack. Place your feet at around shoulder width, with your toes pointed out at a 30- to 45-degree angle.
2. Keeping your chest up and your back flat, push your knees out as you bend them. Focus on pushing your knees out and forward over your toes as far as possible (so long as you maintain heel contact with the ground and can generate force through your whole foot).
3. Stay as upright as possible by maintaining a flat back and holding your chest high. Descend to get your thighs deep as you can below parallel to the ground while maintaining a flat (neutral or lordotic, not kyphotic) back.
4. Hold that bottom position for a second or two, and then stand straight up by leading with your chest up.

MUSCLES INVOLVED

Primary: Quadriceps (rectus femoris, vastus lateralis, vastus medialis, vastus intermedius), gluteus maximus, gluteus medius, gluteus minimus, adductor longus, adductor magnus, adductor brevis, pectineus, gracilis

Secondary: Soleus, rectus abdominis, erector spinae (iliocostalis, longissimus, spinalis), transversus abdominis

EXERCISE NOTES

As you descend into the squat, you'll be tempted to cave your chest down and forward. By keeping your chest sky high during the whole movement, you can maintain a safer position that allows you to get deep while emphasizing your quads and minimizing axial fatigue. There's no hard rule about foot width or angle, so play with a few options to find a good fit. You can get a good clue as to what will work best by squatting down fully with no bar on your back and just letting your feet and toes move around a bit. You want to find the position that your ankles, knees, and hips find most comfortable, and then remember that position for your warm-up and working sets. Though it's tempting to rush through, especially the bottom third of the eccentric phase should be done slowly to maximize effect and minimize injury risk.

TIMING CONSIDERATIONS

When you get strong enough, perform more isolated quad work (such as leg presses or hack squats) before squatting to lower the amount of load you need to use on this exercise. This will lower the axial and systemic fatigue.

SAFETY

To lower the risk of injury, keep your lower back flat at all times by keeping your chest up.

VARIATIONS

You can try different foot stances of the high bar squat, but also try to do the exercise inside the Smith machine. When doing Smith machine squats, place your feet a few inches (about three to four centimeters) in front of the bar path. Stay as upright as you can so that you can minimize axial fatigue even more while maximizing quadriceps stimulus.

HACK SQUAT

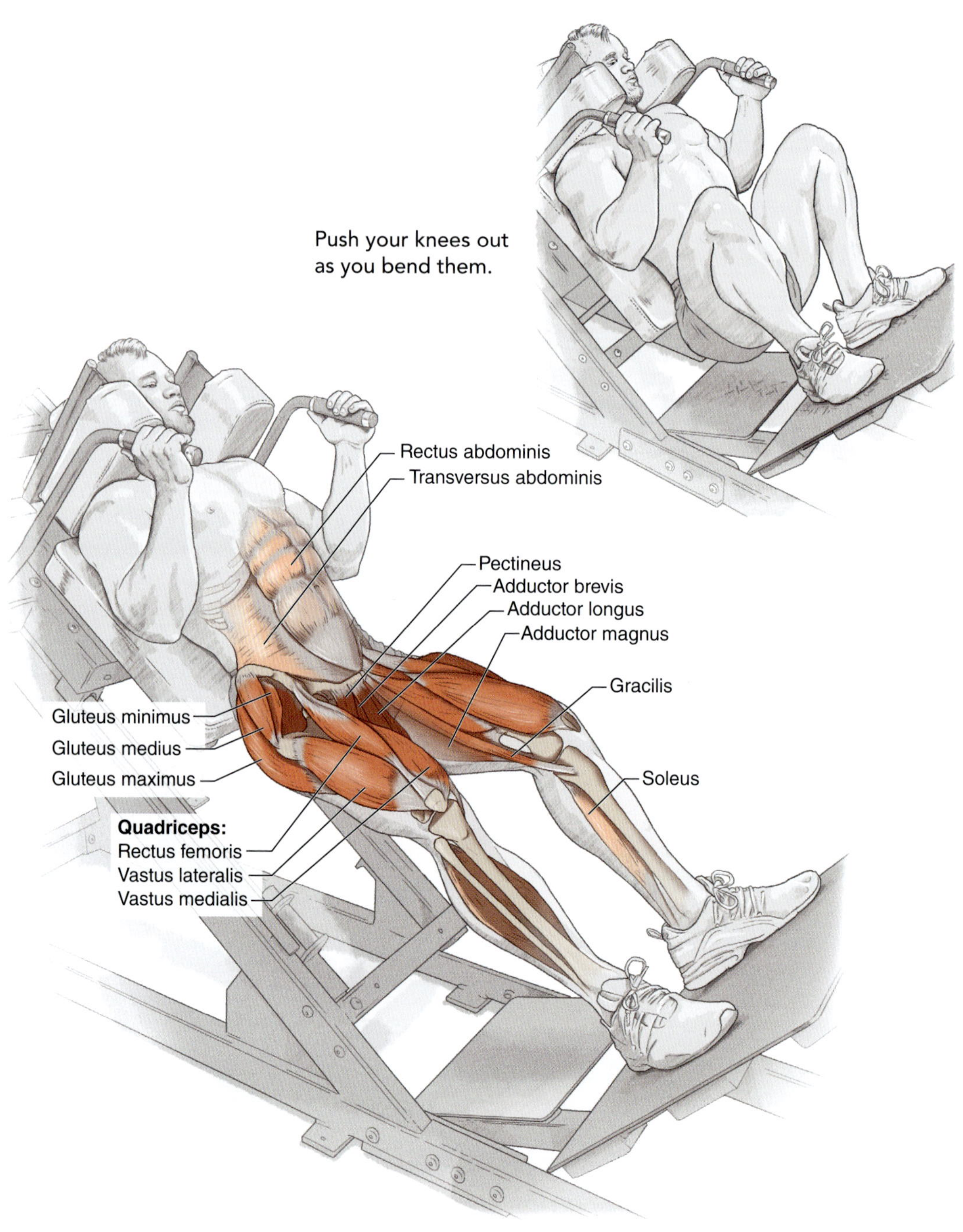

EXECUTION

1. Position yourself in the hack squat machine, with your feet just inside shoulder-width apart and your toes pointed out at a 30- to 45-degree angle.

2. Keeping your chest up and your back flat, push your knees out as you bend them. Focus on pushing them out and forward over your toes as far as possible (so long as you maintain heel contact with the platform and can generate force through your whole foot).
3. On the descent, stay as upright as possible by maintaining a flat back and with your chest held high. Descend as deep as you can while maintaining a flat (neutral or lordotic, not kyphotic) back.
4. Hold the bottom position for a second or two, and then stand straight up by leading with your chest.

MUSCLES INVOLVED

Primary: Quadriceps (rectus femoris, vastus lateralis, vastus medialis, vastus intermedius), gluteus maximus, gluteus medius, gluteus minimus, adductor longus, adductor magnus, adductor brevis, pectineus, gracilis

Secondary: Soleus, rectus abdominis, erector spinae (iliocostalis, longissimus, spinalis), transversus abdominis

EXERCISE NOTES

The lower you can put your feet on the platform and still push from your heels and whole foot, the better. This involves and stimulates the quads the most. If you can't keep your heels down on warm-ups, abort the rep. Come back up, and move your feet up on the platform an inch or two (four centimeters) so you can try again to see if your heels stay down. Perform this exercise with a slow eccentric phase to try to generate a lot of growth (and pain). Keeping your chest up and big at the bottom can reinforce proper positioning for your back and lower body and further enhance the effectiveness and safety of this exercise.

TIMING CONSIDERATIONS

Intermediate and advanced lifters may benefit the most from prioritizing the hack squat; beginners should primarily focus on barbell exercises to achieve fundamental movement proficiency. Due to the significant growth potential of a few basic exercises, beginners may not require the added variation offered by the hack squat.

SAFETY

Your best bet for safety is to pause at the bottom. Because it extends the duration of the loaded stretch, pausing is usually growth promoting as well.

VARIATIONS

Because you can lock your knees out and rest between reps at the top of this exercise, it's well suited for myo-reps—that is, if you're ready to take your muscle growth and discomfort (and pain and breathlessness) to a new level!

LEG PRESS

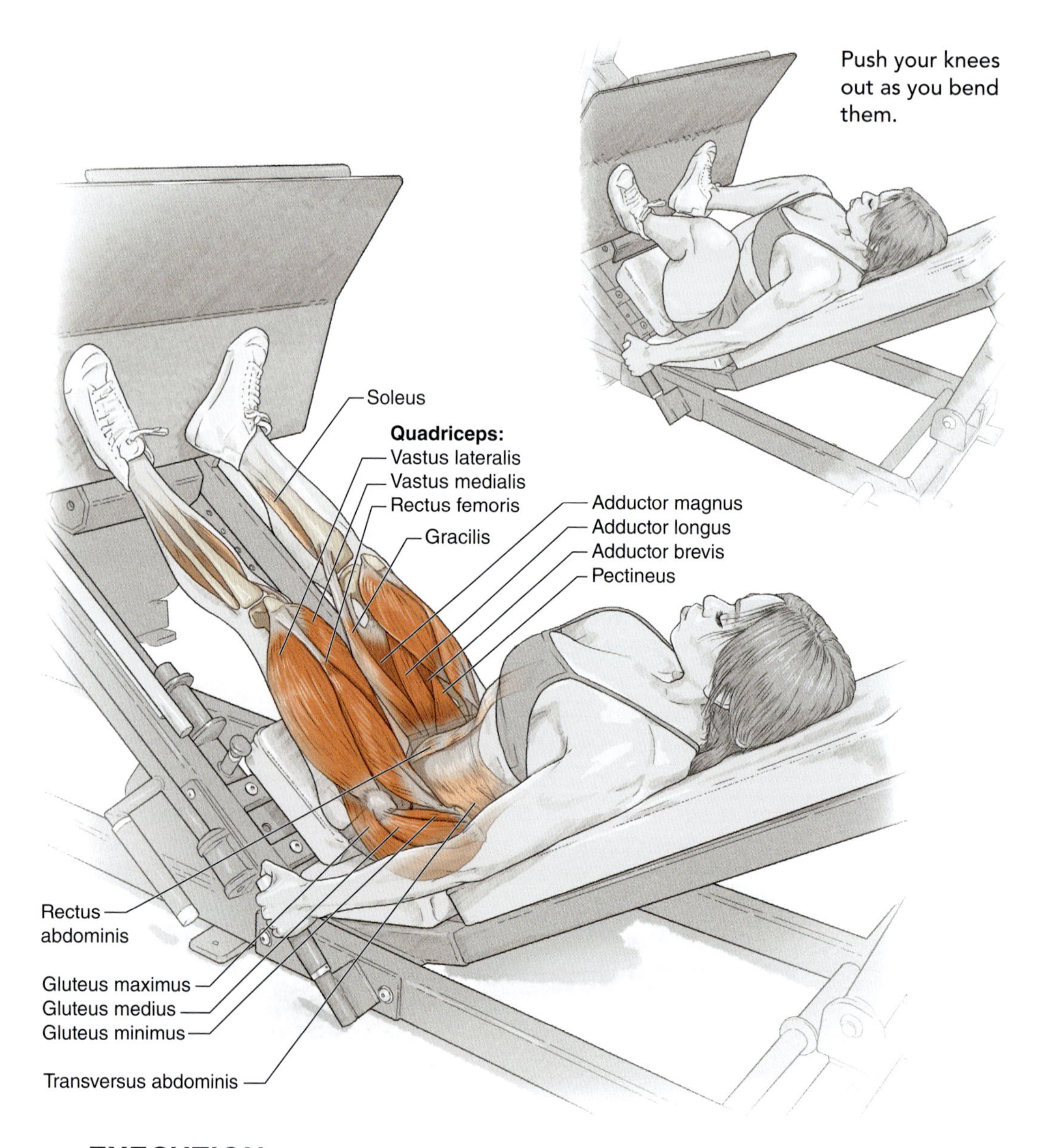

EXECUTION

1. Place your feet on the platform at just inside shoulder-width, with your toes pointed out at a 30- to 45-degree angle.
2. Keeping your chest up and your back flat, push your knees out as you bend them. As your knees bend, focus on pushing them out and forward over your toes as far as possible (so long as you maintain heel contact with the platform and can generate force through your whole foot).
3. On the descent, maintain that flat back, and hold your chest high. Descend as deep as you can while maintaining a flat (neutral or

lordotic, not kyphotic) back; move as though you want your knees to touch your armpits.

4. Hold that bottom position for a second or two. Then push straight up, keeping your chest up.

MUSCLES INVOLVED

Primary: Quadriceps (rectus femoris, vastus lateralis, vastus medialis, vastus intermedius), gluteus maximus, gluteus medius, gluteus minimus, adductor longus, adductor magnus, adductor brevis, pectineus, gracilis

Secondary: Soleus, rectus abdominis, transversus abdominis

EXERCISE NOTES

To optimize quad stimulation during this exercise, place your feet on the platform at the lowest possible position, while maintaining pressure on your heels and the whole foot. If you are unable to keep your heels down during warm-ups, we advise aborting the repetition. Move your feet up on the platform by one or two inches (about four centimeters), and attempt the exercise again to determine if your heels can remain in contact with the platform. Employing a slow eccentric phase is another effective strategy to promote muscle growth, although it may also produce significant discomfort. Additionally, maintaining a large and upright chest position at the bottom of the movement reinforces proper alignment of the lower body and back, thereby enhancing both the effectiveness and safety of the exercise. When you think you've gone as low as you can, try to go a little lower while keeping your lower back flat. You may find that the last little bit of depth pays disproportionately high dividends on the benefits of this exercise.

TIMING CONSIDERATIONS

Intermediate and advanced lifters may benefit more from focusing on the leg press, while beginners should primarily concentrate on basic barbell exercises to develop fundamental skills. Additionally, beginners often experience significant muscle growth from a few core exercises, making variations unnecessary.

SAFETY

As long as your lower back is tight, you're good to go. Do not round your lower back at the bottom, especially if you can feel notable pain or discomfort while doing it. It is safe to lock out your knees at the top, unless you're doing this exercise with out-of-control explosiveness or have a congenital joint hypermobility disorder.

VARIATIONS

Wider and narrower stances can be effective, so long as your spine, hips, knees, and ankles are comfortable. If you're using a wide stance, make sure to simultaneously turn your toes out, so that your toes point at least as far out as your knees do.

PENDULUM SQUAT

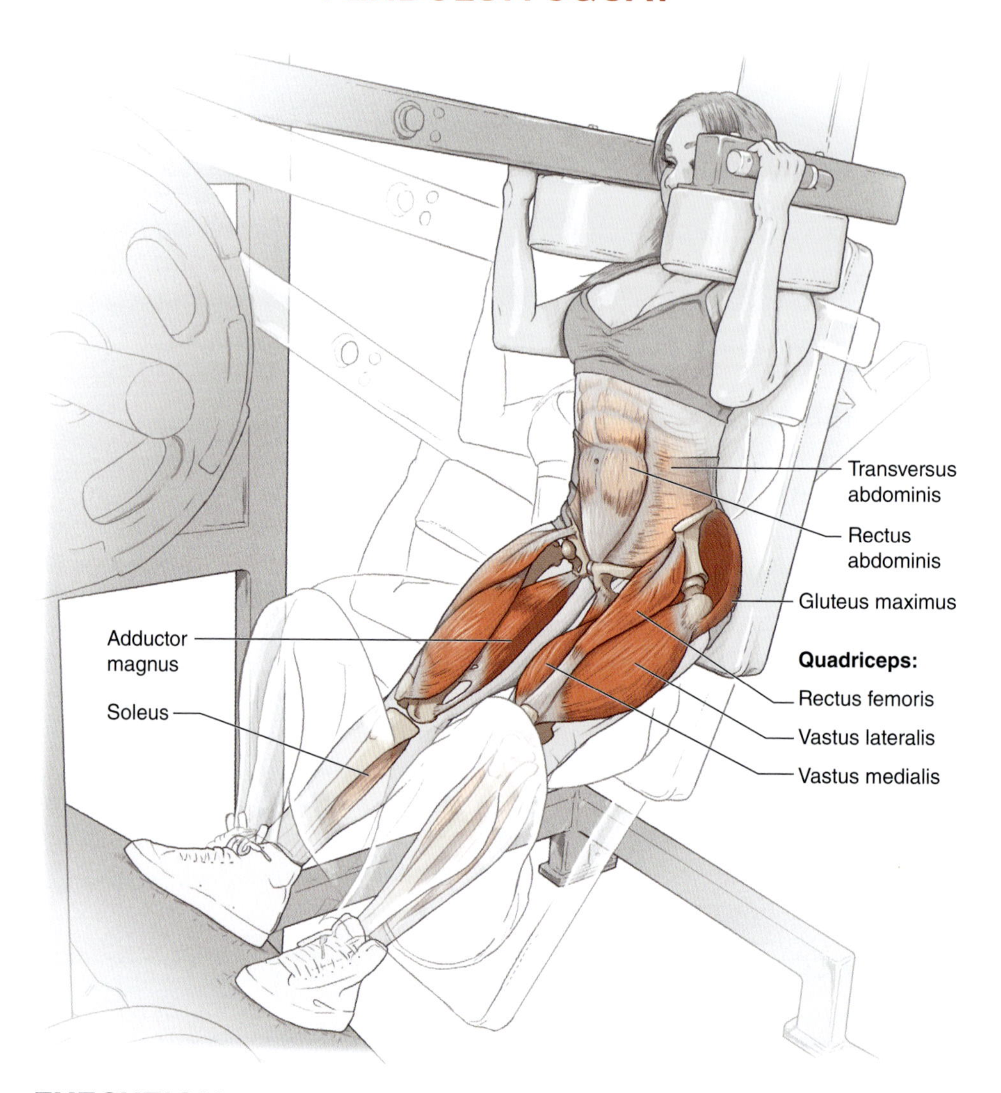

EXECUTION

1. Position yourself in the pendulum squat machine, with your feet placed just inside shoulder-width apart and your toes pointed out at a 30- to 45-degree angle.
2. To perform the exercise, bend your knees while keeping your chest up and your back flat. Focus on pushing your knees out and forward over your toes as far as possible, while maintaining heel contact with the platform.
3. As you descend, try to stay as upright as possible by keeping your back flat and your chest held high. Lower yourself as deeply as you can while still maintaining a neutral back.
4. Hold this bottom position for a moment or two. Then stand back up by leading with your chest. Throughout the descent, focus on maintaining a controlled eccentric contraction and generating force through your whole foot to ensure maximum effectiveness.

MUSCLES INVOLVED

Primary: Quadriceps (rectus femoris, vastus lateralis, vastus medialis, vastus intermedius), gluteus maximus, gluteus medius, gluteus minimus, adductor longus, adductor magnus, adductor brevis, pectineus, gracilis

Secondary: Soleus, rectus abdominis, erector spinae (iliocostalis, longissimus, spinalis), transversus abdominis

EXERCISE NOTES

If you place your feet high on the platform, you can lift more weight; however, that will come at the expense of the best possible quad stimulus and also force you to round your lower back (potentially increasing your risk of injury to that area). Instead, place your feet as low on the platform as you can, as long as you can still drive with your heels and whole foot at the bottom. Because this exercise is leveraged to dump you down through the eccentric phase as quickly and forcefully as possible, resisting and slowing the eccentric phase from the top (and especially through the middle) of the movement confers both more hypertrophic benefit and more safety. Get ready, though, because the onset speed and force of the eccentric is intense in this exercise—it may catch you off guard if you're not prepared to slow down the load with your quads!

TIMING CONSIDERATIONS

Because of the unusually intense eccentric midrange loading on this exercise, it's great for building strength and confidence for leg presses and hack squats later; it helps foster a sensation of stable and strong knees and lower quads. If you find you're able to control your pendulum squats, then leg presses and hack squats will feel very easy to control on the descent (by comparison). You will likely be able to execute those exercises with even more effort and effect.

SAFETY

A rapid eccentric contraction during this exercise is inadvisable, as it can result in a forceful impact on landing, generating excessively high forces at a fast velocity of muscle lengthening. Such an impact increases the risk of injury, particularly if you fail to control the eccentric phase. It's also advised to stop at most one rep short of failure on this exercise; true failure will pin you in the machine with no obvious, easy, safe, or comfortable way out.

VARIATIONS

This exercise is best suited for straight sets in most cases, but myo-reps can also work (especially if you're good about keeping up with your breathing during the later reps).

ROGERS SQUAT

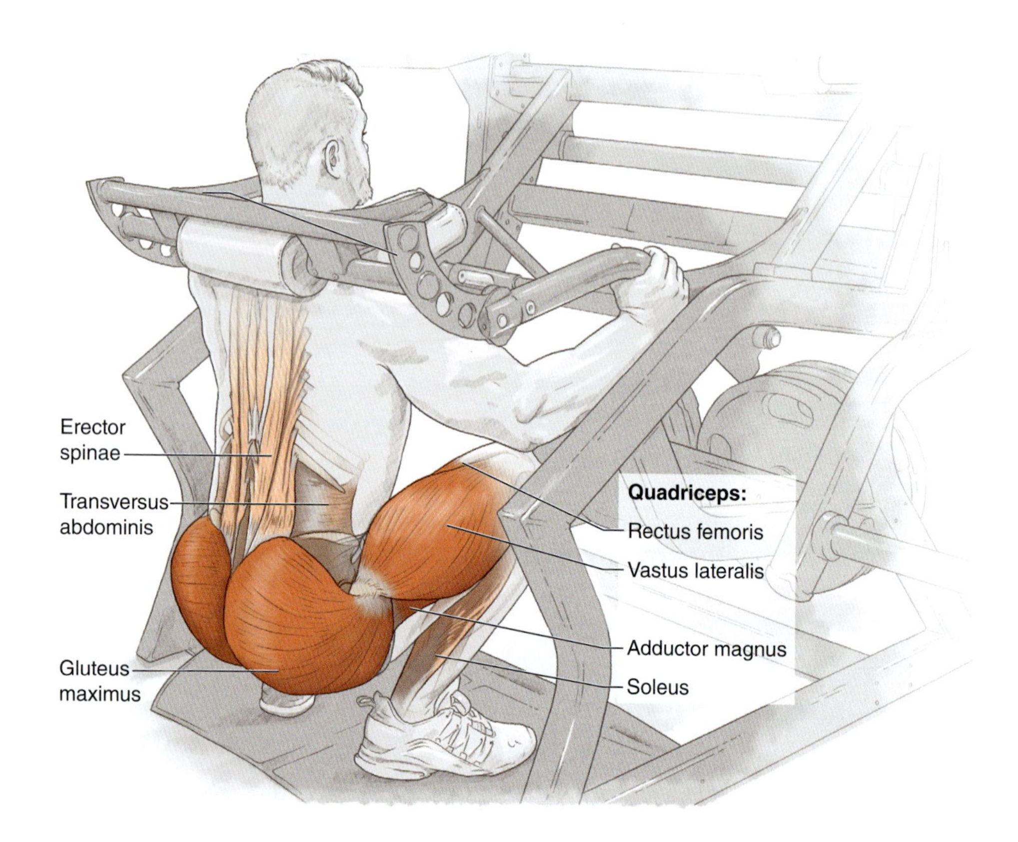

EXECUTION

1. Position yourself in the Rogers squat machine. Put your feet just inside shoulder-width, toes pointing out a bit.
2. Bend your knees while keeping your chest up and your back flat. Focus on pushing your knees out and forward over your toes as far as possible, while maintaining heel contact with the platform.
3. As you descend, try to stay as upright as possible by keeping your back flat and your chest held high. Lower yourself as deeply as you can while still maintaining a neutral back, maintaining a controlled eccentric contraction, and generating force through your whole foot to ensure maximum effectiveness.
4. Hold this bottom position for a second or two. Then stand back up by leading with your chest.

MUSCLES INVOLVED

Primary: Quadriceps (rectus femoris, vastus lateralis, vastus medialis, vastus intermedius), gluteus maximus, gluteus medius, gluteus minimus, adductor longus, adductor magnus, adductor brevis, pectineus, gracilis

Secondary: Soleus, rectus abdominis, erector spinae (iliocostalis, longissimus, spinalis), transversus abdominis

EXERCISE NOTES

Generally there are two ways to position plates for the load when setting up a Rogers squat. One of the options will leverage the weight to be roughly parallel to the ground at the bottom, and the other will have the weight at a considerably higher angle. Most or all of the load should usually go on the option that places the load at parallel to the ground in muscle growth training, because this means that the load lifted by the lifter will be maximized at the bottom. The literature confirms that emphasizing bottom-loading grows more muscle. You can use the other set of loading pegs, but those are better for training explosiveness for sport applications (rather than for muscle growth purposes).

TIMING CONSIDERATIONS

If you place your feet out a bit in front of you, you can greatly reduce the emphasis on the back and posterior chain. That makes this exercise a great choice for quad training if you've already trained your hamstrings in that session with hinges such as the SLDL and good morning exercises.

SAFETY

If you go deep and experience a maximum quad stretch at the bottom, you're doing the exercise correctly. If this is not occurring and you bottom the machine out before going to full depth, you can place weight plates or stable mats under your feet to increase the depth. At the very bottom, it will be the hardest spot for you and a position you can't really get out of; if you're close to failure, consider a spotter to help you out of that last rep.

VARIATIONS

This machine is well suited to myo-reps because of the low amount of axial fatigue; it offers the ability to reduce acute fatigue when standing with knees locked out, resting between myo-rep set intervals. Because racking and unracking are easy, you can even unrack, take 10 to 15 seconds for a few breaths, and go again. Keep breathing through every rep; don't hold your breath except during the last bottom third of the movement, or you will be limited by your cardio before you're limited by your quads.

SINGLE-LEG STAIR CALVES

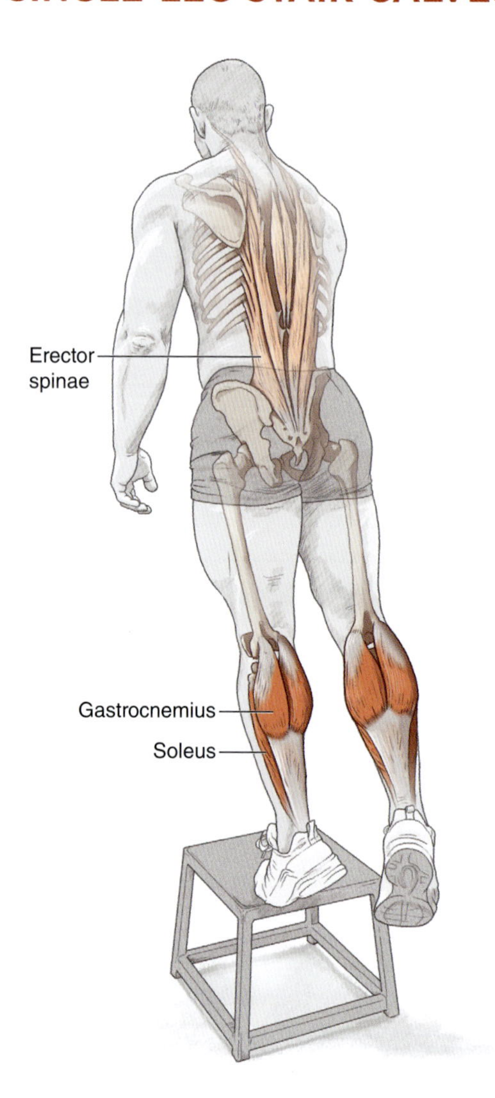

EXECUTION

1. With a dumbbell in one hand, stand on a step. Grab a solid object to stabilize yourself with the other hand (a step railing is ideal).
2. Curl one leg so the foot doesn't touch the step. Shift the other foot halfway over the ledge of the stair. Slowly descend with control, and push your knee back while descending.
3. When you can't descend any lower, hold that bottom position for one to two seconds. Then extend back up. You can extend to a full muscle contraction; however, extending just enough so that your heel is above the stair you're standing on may offer even more growth because it will bias the movement to a lengthened position.
4. As you begin to descend for the next rep, make sure you go slowly on that descent (just as you did with the prior rep). When you finish a

set, rest for 5 or 10 seconds to catch your breath. Then switch hands with the dumbbell and stair railing, switch to the other foot, and begin your next set.

MUSCLES INVOLVED

Primary: Gastrocnemius, soleus

Secondary: Forearm muscles, rectus abdominis, erector spinae (iliocostalis, longissimus, spinalis)

EXERCISE NOTES

Comfort and purchase are very important with calf raises, especially when moving off one leg at a time. Play with your foot position, so that the ball of your foot is on the edge of the stair you're standing on, but not over it. This will give your foot maximum purchase and force-generating ability; it will also minimize discomfort from the edge of the stair rubbing into the foot. As you do the reps in each set, your foot might slip from its original best position. If that occurs (and it probably will regularly), take a second break from doing reps to readjust your foot position before you continue.

TIMING CONSIDERATIONS

Because you have to hold a dumbbell in one hand—and because the stability of this movement is low (one leg at a time, foot slipping over the course of multiple reps)—higher reps using lighter loads are likely best here. Your first set should involve about 20 to 30 reps, and (due to the short rest between sets) your latter set should involve about 5 to 15 reps in most cases. It's likely best to rest for only a short amount on this exercise, as this exercise is low in systemic and cardiovascular fatigue imposition; just waiting for the other leg to finish gives the unused leg more than enough time to recover for another productive set.

SAFETY

This exercise is safe, but be attentive to your foot so you can adjust and avoid slipping. Proper footwear is key here; tight, close-toed shoes will be vastly superior to slip-ons or bare feet. And for those of us who like to train in Crocs™, this is as true as it is difficult to accept!

VARIATIONS

The sky is the limit with variations on this exercise. You can do drop sets with no weight, placing the dumbbell down on the stair in front of you as soon as you near failure and then continuing the set without it. You can do myo-rep sets or myo-rep match sets, and even giant sets work if you prefer to just crank out a total number of reps over many sets. Technically, you can also use both feet at the same time on this exercise; however, for many people, gaining strength will result in having to hold dumbbells that are unwieldy. If your grip is a limiting factor on your ability to hold the dumbbell, feel free to use Versa Gripps or straps.

SEATED STRAIGHT LEG CALF RAISE

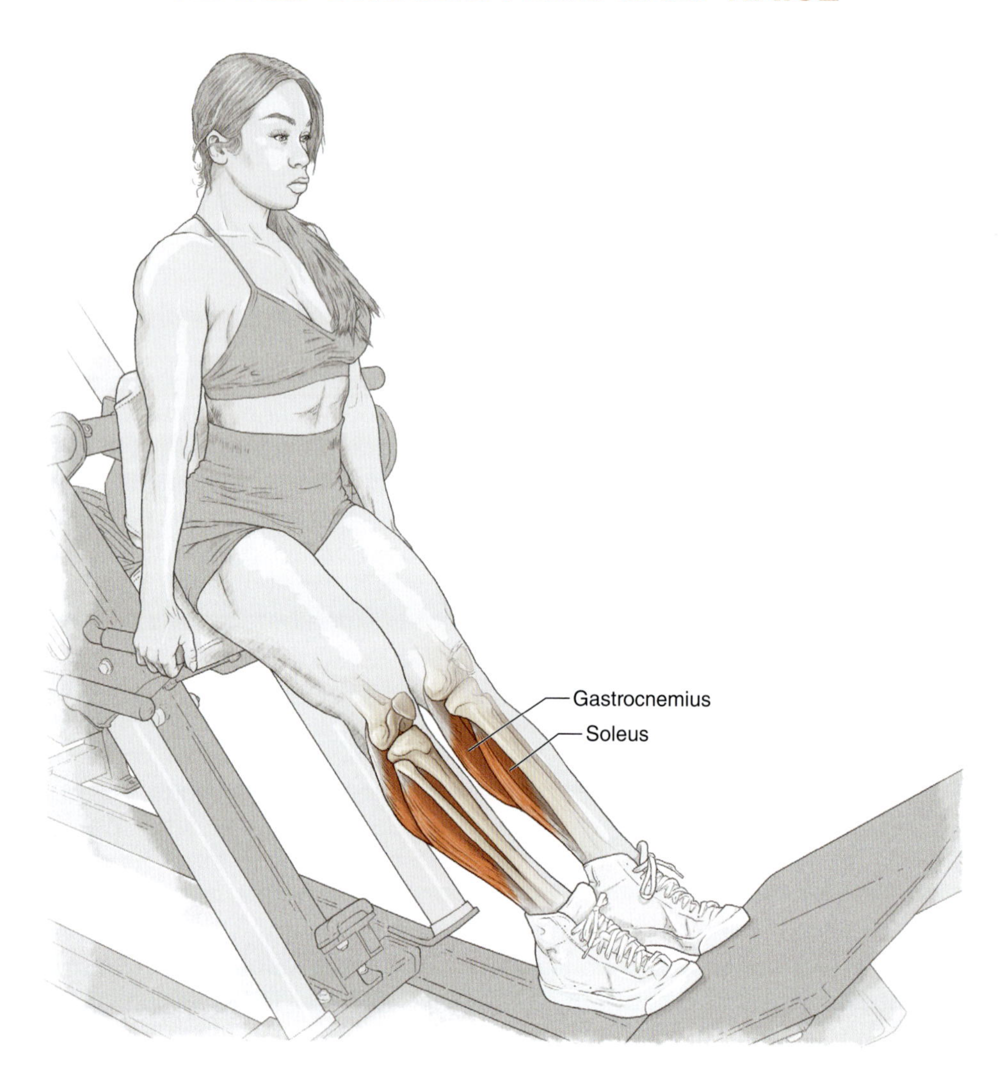

EXECUTION

1. Place the balls of your feet just on the inside of the platform edge.
2. Slowly descend to a full stretch at the bottom, keeping your knees completely locked out so that the gastrocnemius is maximally stretched.
3. Hold for one to two seconds at the bottom. Then quickly come back up to the top. You don't have to hold the top position; in fact, you can come up just high enough so that your heels are just a bit above parallel with the edge of the platform before you begin your next rep's slow descent.

MUSCLES INVOLVED

Primary: Gastrocnemius, soleus

EXERCISE NOTES

Optimal comfort and stability play a pivotal role in performing machine calf raises. It is critical to experiment with the alignment of your foot, ensuring that the metatarsal region (the ball of your foot) is accurately placed on the utmost boundary of the stair tread, without exceeding it. This optimal placement enhances traction and subsequently boosts force production capacity, while minimizing any potential discomfort resulting from the stair edge's abrasive contact with your foot. During the execution of individual repetitions within each set, your foot may displace from its ideal position. This may occur frequently; each time it happens, take a periodic momentary pause from your repetitions to realign your foot before you proceed.

TIMING CONSIDERATIONS

Because you have maximum stability and control on this exercise, you can go a bit heavier here, with sets of as few as 5 to 10 reps. That being said, you need to make sure you're still descending slowly and holding the bottom position for maximum safety and effectiveness. Be honest about which range of reps gives you the biggest pumps, the most local muscle fatigue, and the most soreness; this can often occur around 15 reps or more. Avoid lifting as much calf load as you can; it won't impress anyone, and it will give you suboptimal gains.

SAFETY

If you use very high loads and bounce out from the bottom, you're asking for trouble. Also, do not spend only a small amount of time in the deepest-stretched position, as that position is likely the most muscle growth promoting.

VARIATIONS

Because you only need about 5 to 10 seconds to fully recharge the muscle for another productive set on this exercise, you can knock out the sets quickly (so long as each subsequent set has at least five repetitions per set). Drop sets, myo-reps, and down sets are all viable choices. You can also try altering the foot position every few mesocycles, switching between a wider stance with toes pointed out more and a narrower stance with toes pointed straighter.

CHAPTER 7

Choosing Your Exercises

When deciding what exercises are best for you, you have options: You can create a program with exercises that are familiar to you, choose exercises at random, or just try exercises that arouse your curiosity. However, as you become more advanced, these may not be the best approaches. It may be more helpful to your long-term progress to choose exercises more methodically. By being attentive to the process, you should be able to make exercise selections that are much better than those chosen at random. It is well understood that there is an easy, step-by-step algorithm for choosing the exercises in your next program (mesocycle). If you follow this algorithm, you can be confident you are making excellent exercise choices and maximizing your results from the hours of toil you spend in the gym. Later in this chapter we'll give you that exact algorithm and teach you how to use it—but first, let's get on the same page with the basic concepts that underlie the operation of the algorithm.

Stimulus-to-Fatigue Ratio (SFR)

The stimulus-to-fatigue ratio (SFR) represents a fundamental concept rooted in both empirical evidence and theoretical support within the realm of hypertrophy training. It acknowledges a well-established observation pertaining to the exercises employed in this training. In essence, each exercise in your repertoire contributes to a specific magnitude of growth stimulus. Concurrently, however, every exercise exacts a toll on your body, in the form of fatigue. The overarching objective lies in maximizing the stimulus for growth while simultaneously minimizing the associated cost in terms of induced fatigue. Fatigue isn't a big deal if you're training only one time or a few times a week. The more often we train and the more consistently we train, the more fatigue climbs and accumulates over the weeks (and interferes with our progress in various ways). For any given level of stimulus an exercise gives us, you should aim to minimize the fatigue it creates. In most cases, exercises that drive a ton of stimulus but cause relatively minor fatigue are going to be the best choices in exercise selection. Those types of exercises not only help grow the most muscle in any given unit of time, but they can also be used for longer amounts of time without excessively escalating fatigue—creating more muscle growth in the long term. This is a big deal! It is beyond the scope of this book to discuss the mechanisms of stimulus and fatigue accumulation in depth. Check out *Scientific Principles of Hypertrophy Training* by Mike Israetel, James Hoffman, Melissa Davis, and Jared Feather (2021), which is an excellent guide to such mechanisms. Here, we'll focus on how to roughly proxy how much stimulus and fatigue you're generating from a given exercise or style of training.

Proxying Stimulus

There are at least four stimulus proxies: the mind–muscle connection, pump, perturbation, and disruption.

Mind–Muscle Connection

The *mind–muscle connection* (MMC) is often used interchangeably with *kinesthetic awareness*, but these are not the same concept. If you feel your pecs squeezing at the top of a cable flye, you are in fact kinesthetically aware of the pecs in that exercise. However, this awareness only indicates that a muscle is being used during the movement—it does not indicate that the muscle is being used in a way that will drive a high muscle growth stimulus. You'll need more than just awareness of muscle contraction to define the MMC. You'll need the detection of high tension or burn in the target muscle to make sense of MMC. If you have some measure of both, you have some measure of MMC; the more you have of either, the higher your MMC will be in a given training situation. Let's look at both tension and burn in turn.

Target Muscle Tension

Tension in MMC involves the experience of feeling like your target muscles are producing a huge amount of force relative to their abilities; it also involves feeling like the target muscles are being stretched nearly to their breaking point, especially on the eccentric part of the exercise. If you feel your biceps seemingly tearing (in the good way) during cable biceps curls, it's likely the exercise is highly stimulative of growth—more so than if you feel it mostly in your forearms and the biceps don't feel as tense. Tension is especially likely to be a good proxy for stimulus with lower repetition range exercises, such as sets of five to 15 reps.

Target Muscle Burn

Especially in higher rep exercises (with sets of 15 to 30 or more reps), accumulating metabolites that trigger muscle growth also irritate the nerve endings around your target muscle(s) and cause the burning sensation commonly associated with hard training in the gym. If your target muscle gets a profound burn with higher sets of reps (especially toward the last five or so repetitions), this is a better sign that more growth is being driven than if muscles other than the target muscle(s) are experiencing the burn.

Pump

Because cell swelling is mechanistically and empirically linked to growth, exercises that (for any given number of sets) cause the biggest pumps will more likely stimulate higher levels of growth than exercises that cause less impressive pumps. For example, if leg presses swell your quads to balloon animal status, but hack squats for the same rep range and set number leave your quads mostly deflated, then leg presses are likely (at least on the pump proxy alone) to cause more muscle growth for you at that given time.

Perturbation

Perturbation is a catchall concept. Basically, it means that stimulative muscle contractions that fatigue target muscles cause more growth than contractions that do not cause as much fatigue. Perturbation is composed of three functions: local strength loss, local coordination loss, and increased cramping.

Local Strength Loss

If you can typically curl 100 pounds for 10 reps, you may find that five sets of dumbbell curls later, you can curl 100 pounds for only eight reps. Instead if you did five sets of cable curls and could curl only 100 pounds for four reps, then we can say that cable curls locally fatigue the biceps much more than dumbbell curls (and likely stimulate more growth). This is especially likely if you feel that the biceps are "out of juice" while your shoulders, back, and overall systemic drive feel good and are not the limiting factors on your post-curl test with 100 pounds. You won't often have to test your strength loss, but it will be apparent from simple daily tasks. For example, if you trained your quads for three sets on the leg press machine and then walked down the stairs with no trouble afterward, it is unlikely that such a session was as stimulative of growth as a session of three sets with another leg press machine—three sets that leave you having to hold the hand rails walking down the stairs because your quads have difficulty producing enough force to prevent you from falling!

Local Coordination Loss

Perturbed (and thus likely stimulated muscles) can experience both the fatigue of their local peripheral nerves and excitation-contraction uncoupling at the neuromuscular junction. In other words, perturbed muscles have trouble receiving communication from the nervous system. This is often perceived by the lifter as a reduction in the usually smooth coordination of muscles that target muscle to produce movements of daily life. For example, if your shoulders are so perturbed that picking up your water bottle feels strange and you look clumsy doing it, it's likely that a bigger growth stimulus was delivered than if you did another exercise that left you as capable as ever in coordinating smooth movements after the training session. If you've ever tried to drive after a tough leg or triceps workout, you'll recognize this phenomenon, as it can make pressing the pedals and steering the wheel more challenging than it usually is!

Increased Cramping

In more extreme cases, cramping can result in the severe perturbation of target muscles. While it doesn't occur reliably enough to be used as a dependable stimulus proxy, if your muscles are so perturbed that they are cramping after the workout, there is a higher chance they were stimulated sufficiently, and perhaps even too much. Put another way, if you trained your triceps hard enough to make them cramp after training, it's likely that you stimulated them enough. If one exercise makes your triceps more likely to cramp after training than another exercise, then the first exercise is likely more stimulative.

Disruption

While perturbation can be detected right after the exercise session is over, disruption indicates that the muscle was considerably damaged from the training. If only because high tension and metabolite production cause the most damage, exercises that tend to cause high degrees of damage usually also cause the highest degrees of muscle growth. Proxying damage is indirect (because most muscle damage is imperceptible), but there are at least two ways to tell if a muscle has been notably damaged from training and is, therefore, likely significantly stimulated to grow.

Muscle Tightness

Some muscles don't really get sore no matter how much you train them, but all muscles will feel tight and less flexible than usual for at least some hours after training, if not days. If you're comparing two exercises and, for the same number of sets, one exercise leaves your target muscles feeling tight for some hours or days after training but the other doesn't leave your muscles as tight or they aren't tight at all, then the first exercise is likely stimulating more growth. If the exercises cause a similar level of tightness, but one exercise's tightness lasts for longer, that exercise is probably also the more stimulative one.

Muscle Soreness

While all muscles get tight after training, some also get sore—superficially in a delayed onset fashion, where soreness begins to set in after a few hours and peaks in a day or two. This delayed onset muscle soreness (DOMS) is an indicator that so much damage occurred that the immune system needed to come in to help heal it. Such massive damage is indicative of the muscle having been exposed to a high degree of tension or metabolites, which is predictive of growth. It is important to note that excessive damage, indicated by DOMS that lasts longer than about half a week, can compete with growth resources for mere recovery and lead to less total muscle gain. Thus, we're not looking for the *most* DOMS from an exercise. We're looking for the most DOMS for the fewest number of sets; you should do no more sets than cause moderate DOMS in most cases and thus stimulate the highest levels of muscle growth. For example, if you have one exercise that causes three days of DOMS after two sets, and another exercise that causes a day or so of tightness after four sets, it's likely that the former exercise is more growth stimulative than the latter.

In the next section, we'll talk about how to use these stimulus proxies in an organized fashion to help determine the total stimulus magnitude of a given exercise as compared to others. Before we do that, let's examine the fatigue proxies.

Defining and Proxying Fatigue

In an ideal world, we'd be training in such a way that maximizes stimulus and causes zero fatigue. Thus, we'd be able to train for as many hours as we like and stimulate as much growth as the metabolism would fuel. But in the real world, every rep, set, and workout comes with some fatigue in addition to stimulus. Fatigue can be broken into three subcomponents: local muscular fatigue, joint and connective tissue fatigue, and systemic fatigue.

Local muscular fatigue is the same fatigue that causes perturbation. We don't need to proxy it to determine SFR for two reasons. First, we already have a proxy for it on the stimulus side in perturbation. Second, this kind of fatigue is *stimulative* to muscle growth, or at least highly correlated with stimulus. In other words, we want more of this fatigue, and we don't want to avoid it. If you find yourself saying, "This leg press fatigues my quads too much," remember that is a compliment, not a criticism, of the machine.

For any given exercise, some of the forces imposed are generated and absorbed by the muscles. This is a good thing because muscles are full of tension receptors that use this force to signal muscle growth. Stimulating those receptors, and thus imposing forces on the muscles, is kind of 90% of the reason we're working out to begin with! However, some of the forces are not imposed on the muscles themselves, but rather on the joints and connective tissues. Because joints and connective tissues often take longer to heal from damage imposed by high forces than muscles do, exercises that beat them up the most tend to be the most fatiguing; those exercises cannot be used with as much volume or consistency as exercises that beat up the joints and connective tissues less. We want exercises that stimulate the muscles most and damage the joints and connective tissues least.

Systemic fatigue is the fatigue imposed by exercises onto the biochemical, hormonal, and nervous systems of the *whole body*. When systemic fatigue is high, you are no longer able to maximally contract any of the body's muscles, and overall training quality suffers. In addition, high systemic fatigue—for example, fatigue that involves an increase in catabolic hormones such as cortisol and a decrease in anabolic hormones such as testosterone—can create a less growth-permissive biochemical environment in all the body's muscles. That is definitely something to minimize in your most productive training.

When estimating fatigue for various exercises in training, we can use four proxies:

1. *Joint/connective tissue fatigue:* If one exercise hurts your joints and connective tissue more than another, then the former exercise is likely to be more fatiguing than the latter. This can be true if the pain is acute or if the pain is cumulative, such that over a few weeks of using one exercise versus another, you get more joint/connective tissue pain.

2. *Rating of perceived exertion:* If one exercise is psychologically more taxing than another, then the former is likely contributing more to cumulative, and especially systemic, fatigue. Psychological taxation can reveal itself as higher difficulty in performing the exercise and a higher burden of pain or discomfort during and after the exercise. For example, if pull-ups are pretty easy on your back, not very painful, and don't leave you drained after five sets, they might cause less fatigue than rack deadlifts (which may require you to psych up for them, are highly painful and discomforting during the lift, and leave you drained after).

3. *Unused muscle strength:* A direct measurement of systemic fatigue is its effect on how much strength you lose in muscles that are *not* involved in the exercise you are performing. For example, if after five sets of pull-ups your chest (which was not involved) now has about 90 percent of its usual strength (rep performance), but after five sets of rack deadlifts your chest is only at 70 percent of usual strength, then it's fair to estimate that rack deadlifts cause more systemic fatigue. In that example, rack deadlifts had a greater fatiguing effect on the body's ability to push unused muscles to their limits.

4. *Desire to train:* There is an imprecise (but often observed) relationship between cumulative fatigue and desire to train. This might also work on a local level, where you might not be as excited to train muscle groups that are already fatigued and be much more excited to train fresher muscles. In addition, if you enjoy training, you may experience less psychological fatigue. Because all fatigue is cumulative and affects every other kind of fatigue, choosing

exercises that you enjoy doing more can both detect fatigue and create less of it. If two exercises are roughly equally stimulative, but you just enjoy doing one much more than the other, then the former exercise will likely cause you less fatigue as opposed to the one you enjoy less.

Deriving the SFR

Mathematically, the SFR (being a ratio) is just the stimulus proxies divided by the fatigue proxies.

$$\text{stimulus} = \text{MMC} + \text{pump} + \text{perturbation} + \text{disruption}$$

abbreviated as

$$S = MMC + PU + PE + DI$$

$$\text{fatigue} = \text{joint/connective tissue} + \text{rate of perceived exertion} + \text{unused strength} + \text{desire to train}$$

abbreviated as

$$F = JC + RPE + US + DT$$

Thus, if SFR = S/F, then

$$SFR = (MMC + PU + PE + DI) / (JC + RPE + US + DT)$$

The math is neat when the variables are purely symbolic, but it can help to give them some ordinal values. This can be as simple as assigning a 1 to 3 ranking to the individual stimulus and fatigue proxies: "1" if the stimulus or fatigue proxy is underwhelming compared to most other exercises for that muscle group, "2" if it's about average, and "3" if it's greater when compared to the other exercises. For example, if your MMC is amazing for the cable curl, it might get a "3" on the MMC variable; but if it doesn't cause impressive pumps, it might get a "1" on the pump variable (and so on for stimulus proxies). Regarding fatigue, if the cable curl is super kind to your joints, it might get a "1" (low fatigue) on JC but might be decently tough to grind out and get a "2" on RP (and so on). If you use the three-tier ordinal scale of "low, moderate, high"—and if you are honest—you can quickly rank any two exercises for the same muscle group, run the numbers, and see which has the higher SFR. That's the theory; now let's take it into practice.

Using SFR to Choose Exercises

You can absolutely create spreadsheets to rank SFR and keep track of every single exercise for every single muscle group. But for most people, this would be both excessively laborious and have a low return on investment. A more shorthand method can give you almost all of the benefits of formal SFR calculation and comparison. The shorthand method is also almost as effective as the much more laborious form.

Proxying SFR in the real world often just comes down to judging SFRs informally. For example, if you compare the SFR of two exercises, you can simply be aware of how each exercise feels from the MMC perspective, how big the pumps are, and how gnarly the perturbation and disruption is for a given number of sets. You can also be aware of how tough the exercises are on the joints, how much they drain you systemically, and how much or how little you enjoy doing them. No math—just an awareness of the relative magnitude of those few variables.

Is it possible to derive a workable SFR comparison just like that, by quickly thinking it through in your head? It seems like it's too good to be true, but it works for two reasons: First, the SFR of many exercises are so far apart that it's not difficult to tell when one is notably lower than the other. For example, if you would pay money to do anything but train with a given exercise because you hate it so much, it's probably lower in SFR than the exercise you're excited to do. If an exercise beats up your joints without much of a pump, it's not hard to tell that a different exercise is probably better. Second, if two exercises have a similar SFR, that itself is the only datapoint you need. In other words, if one exercise has a very similar SFR to another, and you're vexed about which one to choose for your next program, it simply doesn't matter which one you choose. If both exercises have a pretty high SFR, you cannot make a wrong choice. Could you try to run the formal comparison method, and determine which exercise is better by a small margin? Yes, you could—but then you wander into another issue. It turns out that because the SFR scale is both subjective and low-resolution ordinal by nature, the estimation error alone completely obfuscates any chance of such a high precision search from revealing any valuable insight. For example, when rating your soreness, you have only a 1 to 3 scale to work with, and your assessment of your soreness is a very subjective process anyway. Therefore, if you compare two exercises on the formal SFR calculation, and one is just a fraction of a whole number higher in SFR, the difference between the two exercises on SFR is well below the reliable precision of the SFR scale (which makes this information pretty useless). Imagine asking an alien who can only discriminate between human heights by a foot and no less to judge if two guys, one who is about 5' 8" and the other about 5' 10" who is taller. Because they are not even close to a foot apart in height, the alien would likely tell you that the two humans are the same height; asking it to be more precise would not be possible.

The imprecision of the SFR scale can sound like bad news, but actually it's great news. All you really need to do is ask yourself which exercise of any two exercises zaps the muscles more while being easy on the joints and psychological fatigue. If you can't tell, then they are close enough in SFR to both be valid exercises for your next plan. If you can tell, then they are far apart enough on SFR for the higher SFR exercise to be almost certainly the correct choice.

Now that you know how to use the SFR scale, your job in training is to try to apply it to the exercises you use. Keep at least a tacit SFR comparison list of the exercises for each muscle group that you use often, so that you can make your best choices in program design. Once you choose an exercise and begin to train with it, you can begin to maximize its SFR to get the most out of it. Let's look at that process next.

Maximizing SFR

Once you've chosen an exercise you think has a decent SFR, or if you're trying a new exercise and you're not sure about its SFR, you should take at least the entire next mesocycle using that exercise to attempt to maximize its SFR. The exception to this diligence is if the exercise clearly and unrelentingly hurts your joints more and more as you do it, even within a given session or between given weeks. If you can't adjust your technique to make the escalating joint pain go away, you should be using a different exercise. If that doesn't occur (and it usually doesn't), you can use the exercise for a whole mesocycle and work to improve its SFR in at least four distinct ways.

Method 1: Modify Tempo

You can try to perform the exercise faster or slower on the way down (the eccentric phase). You can also try to take a pause at the bottom or top of the exercise and see which of these tempos seems to maximize the SFR. If you find that all of them are comparable, that's great because you just found tons of great variations you can use later! For example, if both four-second eccentric reps and one-second eccentric reps have a high SFR for a pull-down, then you really have two pull-down exercises you can program in future mesocycles: a slow eccentric version and a regular version. If, on the other hand, you find that one tempo is superior to the others, you can only use that tempo from now on, knowing that you're using the best tempo possible with that exercise (and thus maximizing its muscle growth yield).

Method 2: Modify Loading Range

Some exercises simply get better SFRs with much lighter or heavier loads. For example, squats of five to 15 reps per set can be amazing for the quads; squats of 15 to 30 reps can become limited by your cardiovascular system and lower back endurance, preventing the quads from being limiting factors and thus lower the SFR of the exercise. On the other hand, dumbbell lateral raises with sets of five to 10 reps can cause a lot of shoulder joint discomfort and minimal delt pumps; dumbbell laterals with sets of 10 to 20 reps can cause a robust pump with no shoulder joint discomfort at all. Before you abandon an exercise as a low-SFR variant, try to use a few different loading ranges to see if one of them is notably superior to the other. Mind you, it's perfectly normal for some exercises to have the same SFR at all loading ranges; that just means more programming variation is available to you later!

Method 3: Modify Technique

If you are training your biceps with a curl, you may find that arcing the bar out in front of you gives you a better MMC for your biceps. Conversely, you may find that drag-curling the bar (or moving it up and down in a straight line) is superior for you at the time. When evaluating an exercise on its SFR, try it with various techniques to see which ones maximize SFR for you. What you may find, as with the above examples, is that you either learn what techniques are better and worse for SFR, or you may learn that a few techniques are similar. Whether you change how much you pull in your elbows during pull-downs or how much your knees travel over your toes in a squat, some exercises have a few different techniques you can try to see which ones have the highest SFRs.

Method 4: Modify Grips and Foot Positions

This change can be small enough to just be your personal preference with a given lift—or large enough to turn that lift into another lift. For example, if you move your hands out one finger-width on the close grip bench press, you're just doing your own version of the close grip bench press; if you move your hands out a whole fist-width, then you're now doing the regular bench press and not the close grip version. Changes small enough to count as variations of the same lift are important to implement to maximize the SFR of any given lift. One of the main reasons this is a wise practice is that we're all built slightly differently. What feels the most comfortable for one person's joints may be a grip width or toe angle that feels a bit off to another person. It's possible that the first time you try a new exercise, it will have an underwhelming SFR. As you play with foot and hand positions to make the lift more comfortable for the joints and stimulative to the muscles, you can notably improve the SFR, allowing the exercise to be considered good enough to keep using often in training.

Deleting and Replacing Exercises

If your chosen exercises all turn out to have great SFRs for the first mesocycle, then you should repeat them in the next mesocycle in most cases. Why break up the band and ruin a good thing? However, SFR will fall at some point for all exercises, whether it occurs after one mesocycle or after 10. You can detect falling SFRs the same way you detect rising ones or stable ones: by doing the shorthand thinking about how your muscles, joints, and psychology are responding. Over a longer timeframe, usually measured in two to four mesocycles, SFRs will fall; this is a phenomenon known as **training staleness**. When we say an exercise is "becoming more stale," that's just means its SFR is declining. For example, you may find that exercises that previously blasted your muscles are no longer yielding the same tensions and pumps. Maybe they aren't getting the muscles as sore as they used to, but the joints are taking bigger hits, and you're kind of just sick of using these exercises. They aren't only getting more boring, but more difficult to muster the inspiration to do them. This happens with nearly all exercises. When it does, we need to figure out one of two things: First, is this exercise, even though it's no longer as high in SFR, still the best choice for our next mesocycle? Or, is the exercise sufficiently low in SFR now, and we could begin using other exercises instead?

By assessing SFR continually, you can better ensure that all of the exercises you're using are your best options at the time. If you've had some exercises in the rotation but their SFRs are not what they used to be, you should at least consider replacing them with higher-SFR variants. We do this exercise deletion and replacement to keep SFR as high as it can be in every program we create. This is because higher SFRs simply lead to greater, more sustainable muscle growth, and that's what training is all about! However, to get the best long-term results, we cannot use SFR alone to determine when to delete and replace exercises. It's 90% or so of the calculus, but at least three caveats apply.

Caveat 1: Conservation of Future Variation and Phase Potentiation

If you have only two chest exercises because you train at home with a pair of dumbbells, you should milk every exercise for all it's got until you switch to the other one, so that each exercise has plenty of time to refresh (in other words, time to decline in staleness and rise in SFR before it's used again). For example, if you used one of the exercises for two mesocycles and you think the other exercise is going to have a slightly higher SFR if you use it next, you can switch. But, if you hold out for another mesocycle and take something like a 10 percent reduction in SFR by using the mildly stale exercise, you might give that other exercise another whole mesocycle to reduce its staleness since you last used it (lifting its SFR by 30 percent for two mesocycles from now and gaining a 20 percent increase in overall SFR). Ideally, you'd switch right away and never have to take that 10 percent hit; but if your total exercise variation is low, especially if you have three or fewer exercise options for a given muscle group, it may be wise to hold off on rapid rotation and squeeze all the decent-SFR training you can out of each exercise before moving onto the next. This way, you avoid the issue of using exercises so often in back-to-back mesocycles that *both* of them become very stale. At that point, it might be time for active rest because you simply have no highly effective exercise to choose from for some time!

Very advanced lifters can concern themselves productively with longer, more complex training phases and more nuanced periodization logic. For example, if you know you have a three-mesocycle block of training after an active rest, you then know that you'll be most sensitive to growth and have the least fatigue in the first mesocycle and be least sensitive to growth and have the most fatigue in the last mesocycle. For simplicity's sake, if you have three exercises with different SFR rankings, you could use the exercise with the lowest SFR first, because you're so sensitive to growth and so fresh that its low SFR is still workable. In the second mesocycle, as you need more stimulus and want to lower fatigue, you could choose the exercise that has an intermediately ranked SFR. Finally, you could save the exercise with the highest SFR for the last mesocycle, precisely when you need both your biggest stimulus and can least afford needlessly large fatigue increases. Imagine if you did it the other way around! If you saved the exercise with the lowest SFR for last, you'd have terrible training precisely when you need to be at your best, and you'd have overkill on effectiveness in the first mesocycle when you just don't need the exercises with a high SFR. Sure, if you had multiple exercises with a high SFR, you could just use those exercises all the time; but if you have limited variation, you can plan your training block intelligently, deploying the exercises with higher SFR when they are most needed.

Caveat 2: Conservation of Performance Momentum

When you're gaining strength in an exercise, able to lift more weight, or able to complete more reps week over week in a mesocycle, this means that you're becoming more efficient with the movement. A part of this efficiency means that you're recruiting higher percentages of the motor units that grow the most—the biggest and fastest-twitch ones. And this means that you're likely getting more growth out of the movement. Thus, separately from SFR, you should consider performance momentum, as seen in your strength progress (in your calculation of whether or not to delete and replace a current exercise or leave it in for the next mesocycle). For example, if your SFR on an exercise is fine but you're not gaining much rep strength on it in this last mesocycle, toss it and replace it. If your SFR is pretty good (not ideal) but you're steadily gaining strength from the exercise, it might be worth waiting until the strength plateaus more or the SFR drops more before replacing it.

Caveat 3: Margin of Error to Current Exercise SFRs

If you replace an exercise that's gaining performance, it is best to choose an exercise that you anticipate will have a *much* higher SFR; you should have a compelling reason to replace an exercise that's gaining performance, such as a much higher SFR. Conversely, it's much more possible that an exercise is stalling in performance but still offers the highest SFR. Sometimes performance stalls or even falls in the last month or two of hard fat loss dieting, and that's not at all a reflection of the exercise choice. *Any* exercise would lose performance in such a catabolic and high fatigue setting. If you replace the stalling exercise with another exercise that has a lower SFR, you will see that one stall as well—while giving you a *worse* muscle growth stimulus! Therefore, if you replace an exercise because of stalling performance, the exercise you are replacing it with should have an SFR at least comparably high. If the SFR of the potential replacement is lower, then there's no need to replace the exercise, no matter its performance characteristics. The main thrust of this caveat is this: If you're going to replace an exercise because performance is stalling, SFR is still the biggest factor. Performance stalling by itself is not enough of a reason to switch exercises. You should be able to note a clear difference in SFR or performance.

You can use the following step-by-step algorithm to determine whether to delete or replace an exercise:

1. Ask yourself if you have the variation to spare a deletion/replacement in this upcoming mesocycle.
2. If the answer is *no*, keep the current exercise.
3. If the answer is *yes, I have a variation*, ask yourself if your performance on the current exercise has stalled.
4. If the answer is *no, my performance has not stalled,* ask yourself if the SFR of the top alternative will be *notably* better than the current exercise's SFR? If the answer is *no*, keep the current exercise. If the answer is *yes*, replace the current exercise.
5. If you answer *yes, my performance on the current exercise has stalled,* ask yourself if the SFR of the top alternative will be the same as or higher than the current exercise's SFR? If the answer is *no*, keep the current exercise. If the answer is *yes*, replace the current exercise.

If you like this algorithm, feel free to use it. If you find that it's too complicated, you can instead remember three things:

1. If an exercise has a great SFR, use it for as long as you can.
2. If a notably better SFR exercise is available and the SFR of your current exercise isn't as good as what it was, consider changing to the new exercise.
3. No matter what exercises you use, try to maximize their SFRs with variations, technique alterations, loading ranges, and tempos.

CHAPTER 8

Posing

The most important factor in building an impressive physique is the training, the diet, and the recovery (during which actual muscular construction happens). But if you'd like to show off your upgraded physique, even if just to check on progress, even if only for yourself, how you present your physique can make a difference. The formal process of presenting various aspects of your physique is called *posing*. While there is definitely an intangible art to it, there are a set of dependable principles that you can use to look your best onstage at a physique show, in gym selfies, and even in your bathroom mirror. There are principles for all of posing, no matter the specific pose, as well as recommendations for each specific pose. We will cover the 11 poses that are mandatory in the Bodybuilding division of physique competition. You do not need to learn how to pose in order to get more muscular and see the benefits visually. However, learning how to pose even a little bit can at least standardize your appearance in every pose, allowing you to more accurately track changes over time (especially as you diet to gain weight or to lose fat).

Universal Truths of Posing

There are a set of universal principles in physique posing that can be applied to almost all poses and posing situations. These principles, when followed, can make every physique look better, and, when ignored, can often render even the most impressive physiques less impactful and aesthetic. Let's discuss these nine principles briefly before we discuss individual poses.

1. *Smile confidently:* Unless you're turned away from the judges, smiling confidently can add to the perceived aesthetic of your physique. By practicing smiling confidently during posing, you can negate the problem of letting your natural facial expressions take over—frankly, many people's expressions look strained and silly!

2. *Open up:* You are posing to show off your muscles, not to hide them. A key principle in nearly all poses is to open up your physique. You can do this by flaring the lats, lifting the chest, not retracting the scapulae, and opening up the knees to show the full size of the legs (inner thighs included). For example, in most poses, having your chest up and out makes you look thicker, wider, and taller at the same time. If you let your chest cave in, even on the rear lat spread pose, you can diminish the impact of your silhouette. If you constrain your physique by curling in, you'll look much smaller than if you open your physique up—and looking as big as possible is kind of a huge deal in posing!

3. *Focus on what's visible:* Do you have to contract your glutes when you're facing the judges in your most muscular pose? Absolutely not. No one will see it, and the extra strain will be visible in your posing and your breathing; it can also tire you out sooner in long posing rounds, degrading your presentation over time. Similarly, if you're doing a rear lat spread with your back to the judges, your tummy will pooch forward if you hit the pose right. Still, because the judges don't see the pooch and because pooching forward allows for your waist to look smaller from behind, all is well! Some novice competitors will flex their entire bodies because they think everything is always on display onstage. But unless some of the judges have X-ray vision, this is not the case.

4. *Contract all visible muscles:* While contracting non-visible muscles in a given pose is unwise, contracting all visible muscles in a pose is a very good idea. This may seem obvious, but countless competitors forget about most of their muscles and focus only on contracting and positioning a few of them. For example, novices tend to contract their biceps and triceps for the front double biceps pose, and some remember to control their midsections, but many forget to contract their legs. It's often said that you should contract the needed muscles from your feet on up, making sure not to miss any.

5. *Employ shallow breathing:* If you breathe with large tidal volumes (huge breaths in and out), you'll experience two negative results when posing: First, you'll end up having a huge, bulging gut about half of the time that you're hitting a pose—neither judges nor fans and photographers will enjoy that look. Second, you'll also illustrate to the judges that you're out of cardio shape for posing, and that's the opposite of both a professional look and the kind of look that ups your placement in the final tally. When onstage, you need to mostly breathe with shallow breaths so that your midsection doesn't move much and stays under your control. Is this tough? Yes. Is it possible to master? Absolutely. It takes many practice sessions of posing to master this, especially longer sessions where you're running through all the poses multiple times and holding each pose a few seconds longer (about five to 10 seconds total) than will usually be asked of you in competition. Judges have very keen eyes for who's in shape to hit the poses like a pro; they definitely know who's slipping up, unprepared.

6. *Don't over-contract:* Contracting your muscles (also called "flexing") is a critical part of displaying them properly. However, if you contract them with as much muscle force as possible, in most cases they won't look appreciably different than if you just contract them the minimal amount to make them taut. In fact, if you contract them any more than this minimal amount, they have a much higher chance of cramping onstage (really bad news), and each pose will tire you out much more, degrading your posing quality as the comparisons proceed. Also, it will be easier to smile and breathe if you aren't trying to contract so hard as to rip muscle from bone. Why do professional bodybuilders often make posing look effortless? Other than years of practice, it's because they're not actually working *that hard*!

7. *Don't shrug up; shrug down:* Shrugging up brings your shoulder joint closer to your midline and makes you look narrower. Shrugging down reverses this path, and makes you look wider. Practice shrugging down in nearly every pose. This is a very common mistake, right up there with over-contracting.

8. *Control your midsection:* Don't just control your midsection with your breath, but also keep your abdominal muscles contracted (barely, but taut) nearly the entire time you're onstage. You can, in fact, relax them during poses where you face away from the judges. For any forward or side-facing pose, make the waist narrow via abdominal control (that also applies to many visible transitions between poses). Novices will often forget to control their waists, and this can make poses like the side triceps pose feature a comical looking, though muscular, pooch belly. This isn't the look you're going for—we promise!

9. *Position before contraction:* Position your limbs where they're supposed to be first. Only after they're in the right place should you contract the needed muscles, in most cases. For example, during the front double biceps pose, arrange your legs as needed, then your core, then flare out your lats, then raise your arms up and get the right shoulder and elbow position, and *only then* contract everything with the needed intensity to present it best. If you go muscle by muscle from the feet up, the same rule applies: Position the limb how you want it, and then contract, repeating all the way up. It's tough to move into positions while contracted, and it also wastes needless energy in high amounts.

Tips for the Bodybuilding Poses

While other divisions (Men's Physique and Female Bikini, for example) have a totally different set of poses, it's a bit outside of the scope of this book to cover them all. Since this book focuses on bodybuilding anatomy, we'll cover the 12 primary bodybuilding poses, giving you some basics to get you started. Almost all bodybuilding competitions will ask their competitors to perform these 12 poses. Each pose is requested, and the "relax" command is announced about three to five seconds after the pose is first hit, with the next pose requested right after. In real competition, most novices are surprised at just how fast the poses are hit in sequence. It turns out that experienced judges are good enough to learn everything they need to rank the athletes from just a few seconds per pose. That being said, sometimes, when two or three athletes are very close in a certain pose or in an overall score, the judges will have you take at least 15 seconds per pose and have you move through a few of the poses a few times. This can be completely exhausting; make it a priority to practice for this without breaking composure.

What follows is a list of all 12 poses and some helpful tips for you to make your own poses the best they can be. Please note, however, that nothing you read here can replace hours of looking at professional posing pictures, videos, and tutorials, and hiring and working with a qualified posing coach. These are just the basics, but they are basics you should learn well!

Front Relaxed

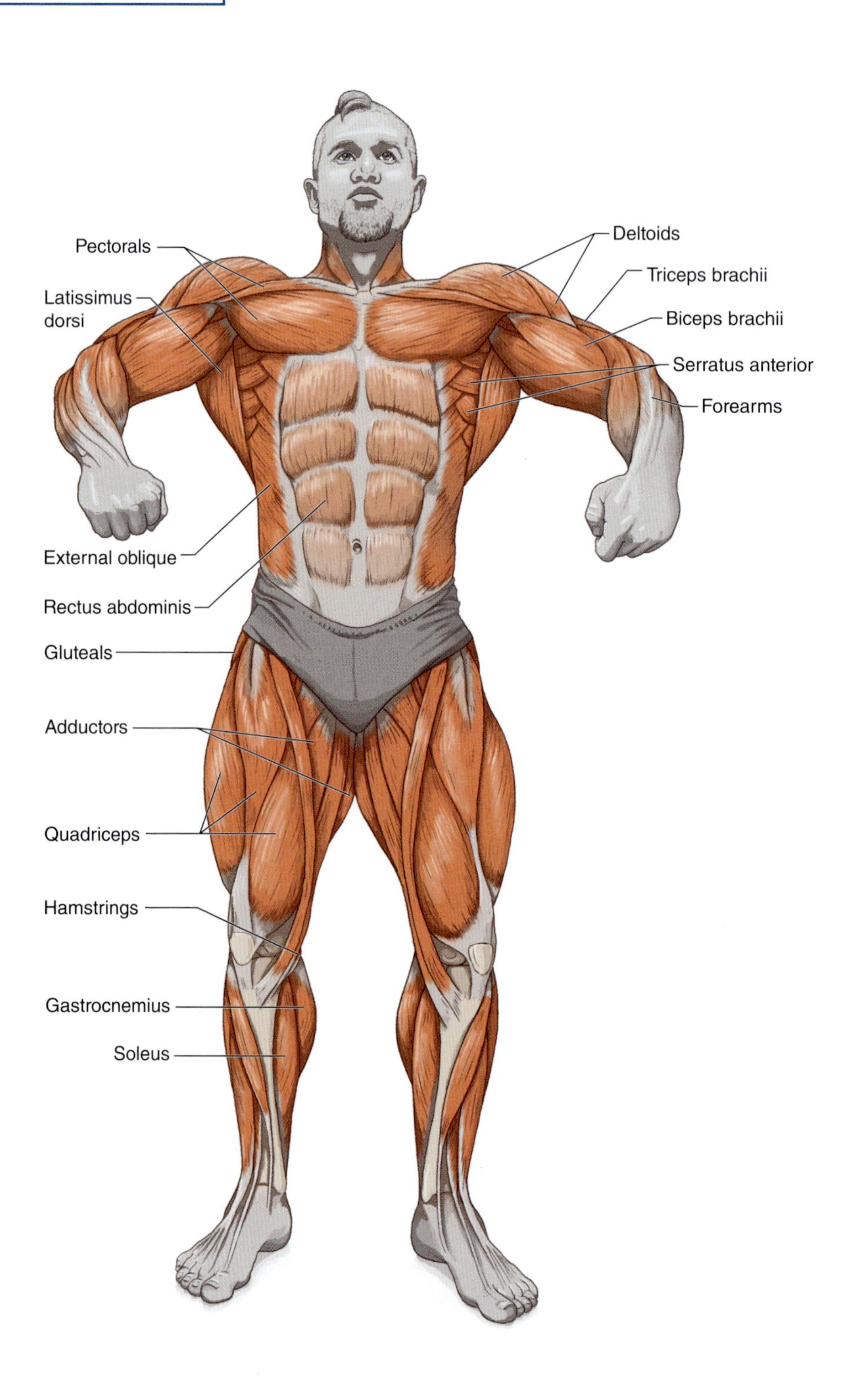

- Shoulders down and out
- Chest up
- 135-degree bend in the elbows
- Lats flared
- Abs in (your best version of a vacuum)
- Weight on the balls of the feet
- Knees unlocked with the legs almost straight
- Quads flared out

Side Relaxed (One from Each Side)

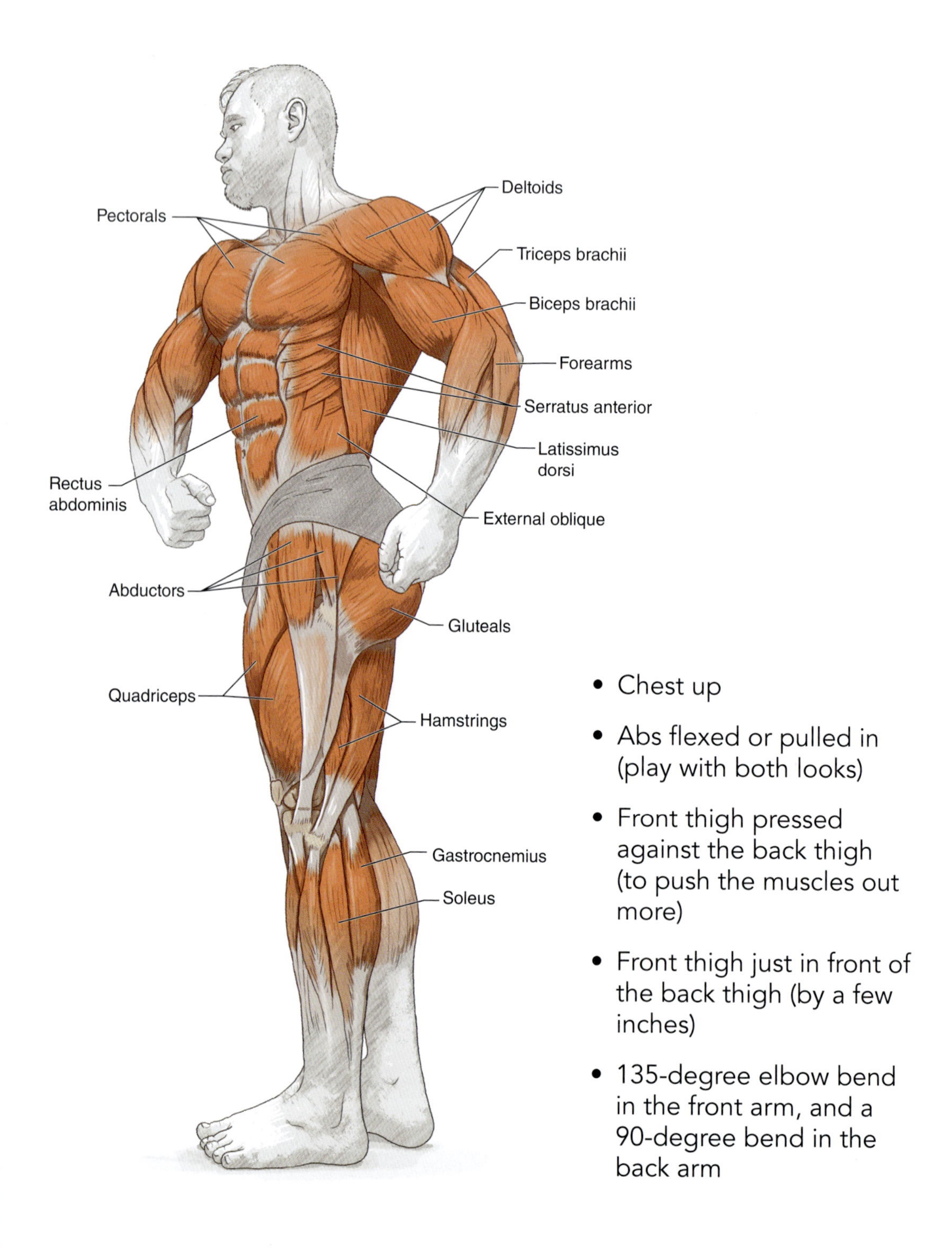

- Chest up
- Abs flexed or pulled in (play with both looks)
- Front thigh pressed against the back thigh (to push the muscles out more)
- Front thigh just in front of the back thigh (by a few inches)
- 135-degree elbow bend in the front arm, and a 90-degree bend in the back arm

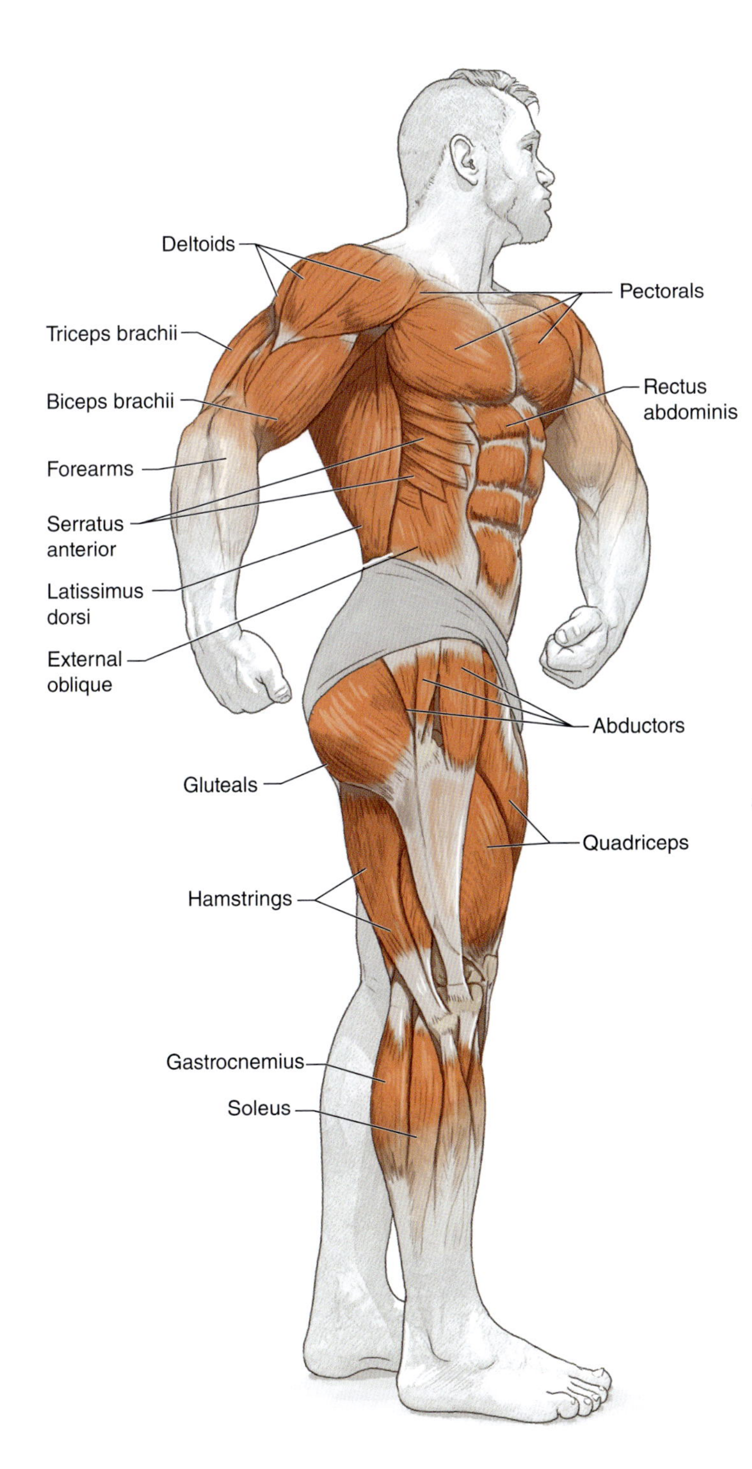

- Front hand behind the glute (not in the way)
- Front shoulder pulled back and back shoulder pulled forward (for illusion of a thicker upper body)

Back Relaxed

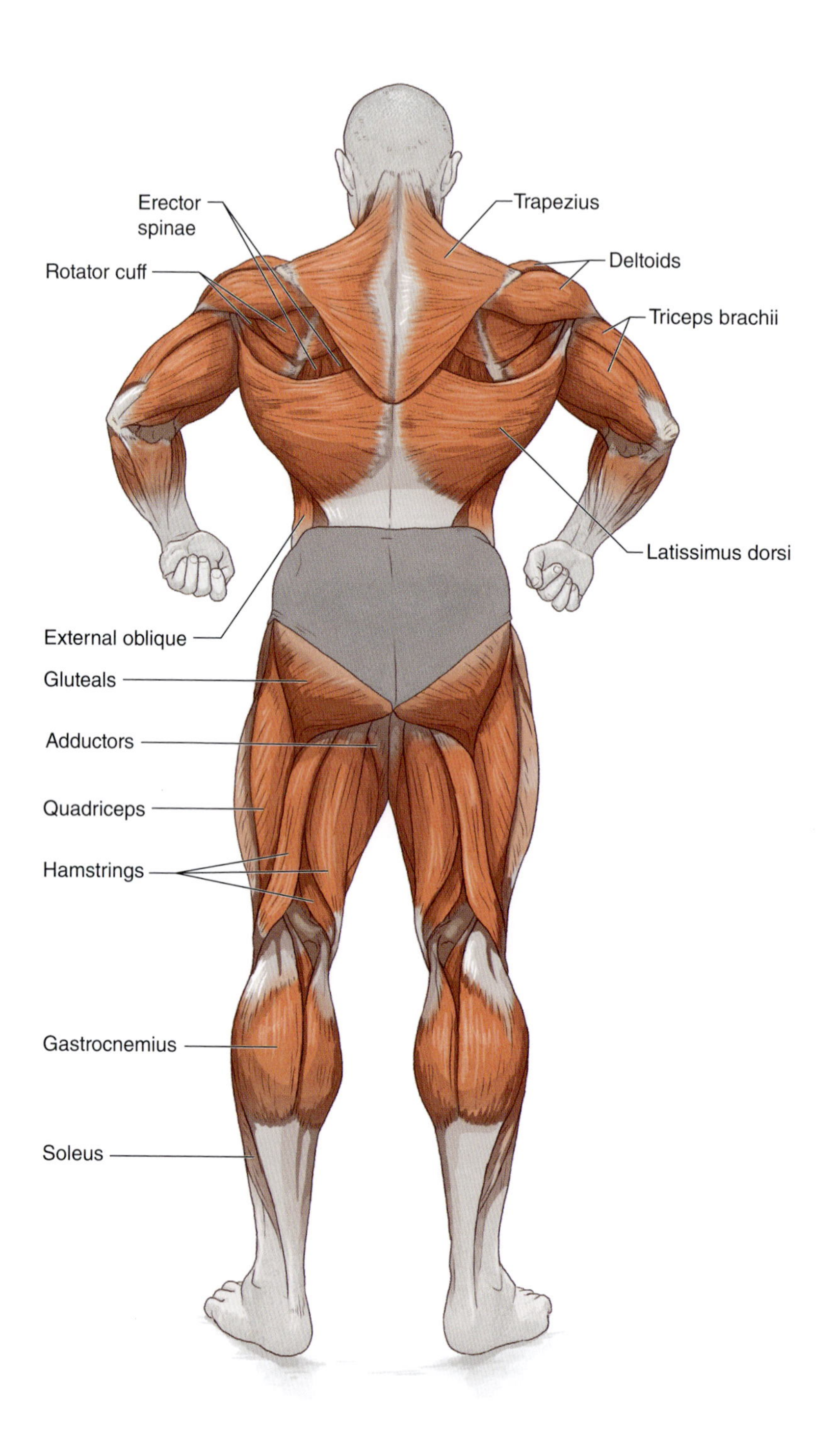

- Quads flared
- Glutes contracted
- Chest up, with a slight lean backward
- Lats flared
- Arms at 135-degree elbow bend
- Scapulae down (never shrugging up)

Front Double Biceps

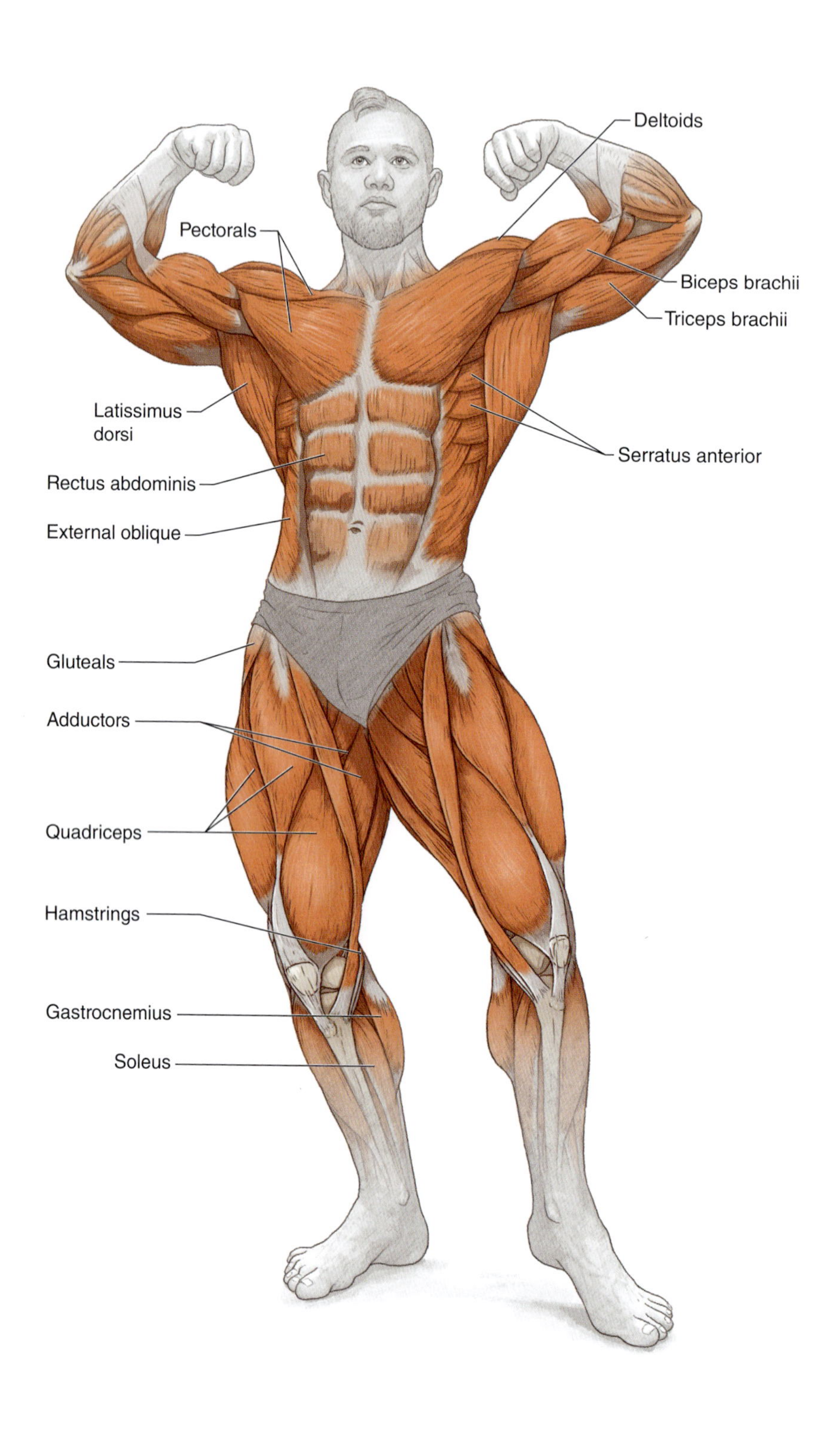

- Quads flared
- Abs in (your best version of a vacuum, though some athletes prefer a crunched down look—try both)
- Chest up
- Shoulders shrugged down
- Lats flared
- Elbows *above the line of your shoulders*
- Elbows bent at 70 to 90 degrees (much less or more is a worse look in most cases)

Front Lat Spread

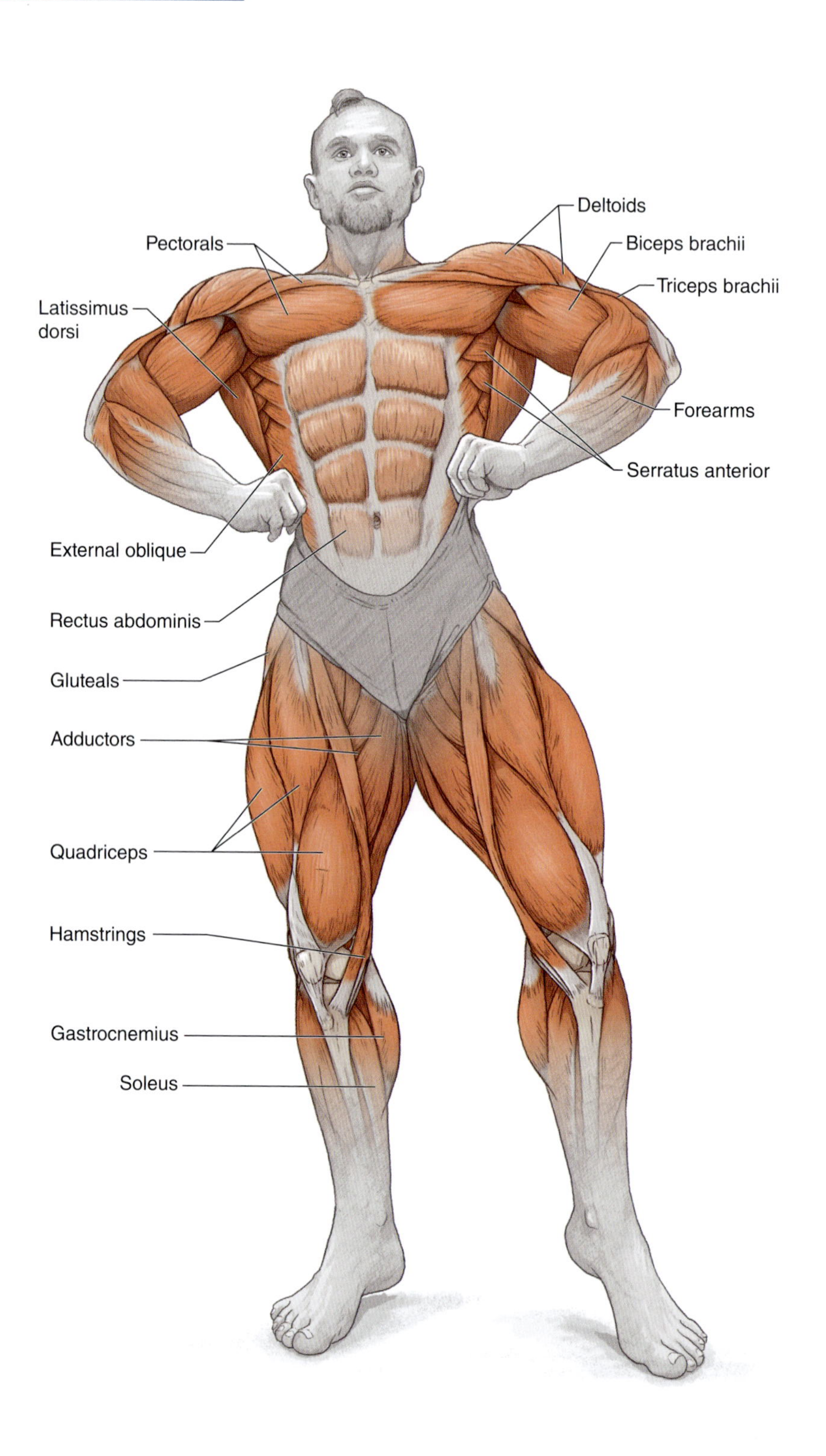

- Quads flared
- Abs in (your best version of a vacuum)
- Chest up
- Shoulders shrugged down
- Elbows out as wide as they can be, directly perpendicular to your midline
- Pinch your waist in with your hands (the exact position of your hands on your waist, and thus your elbow angle, will be dependent on your body shape and needs to be experimented with—start just under your ribs, and go from there)

Side Triceps

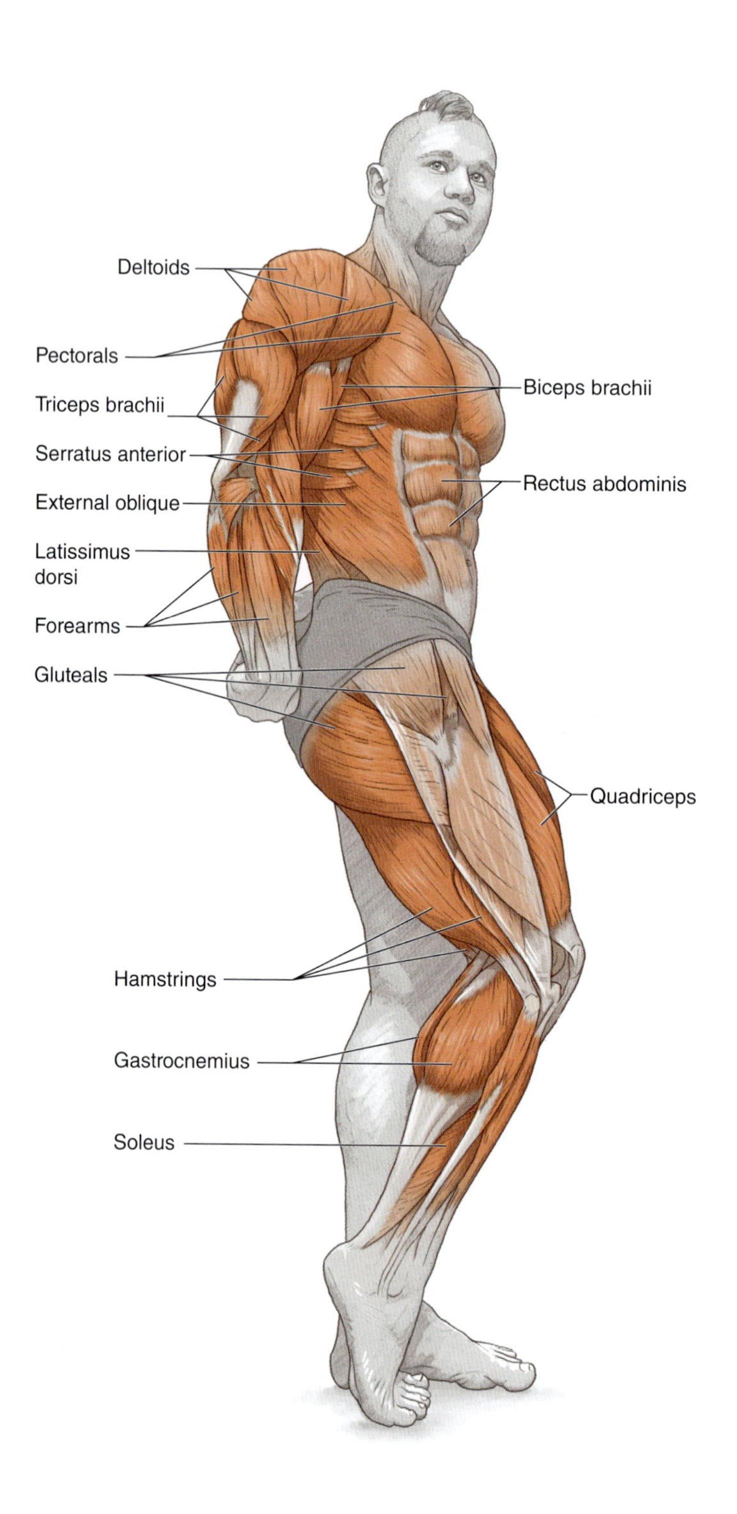

- Front leg about halfway in front of the back leg (some adjustment will be required to get your best look)
- Abs in (either crunched down or vacuum—try both)
- Hips perpendicular to the judges, with the spine rotated as much as possible so that the shoulders are as wide as possible

Side Chest

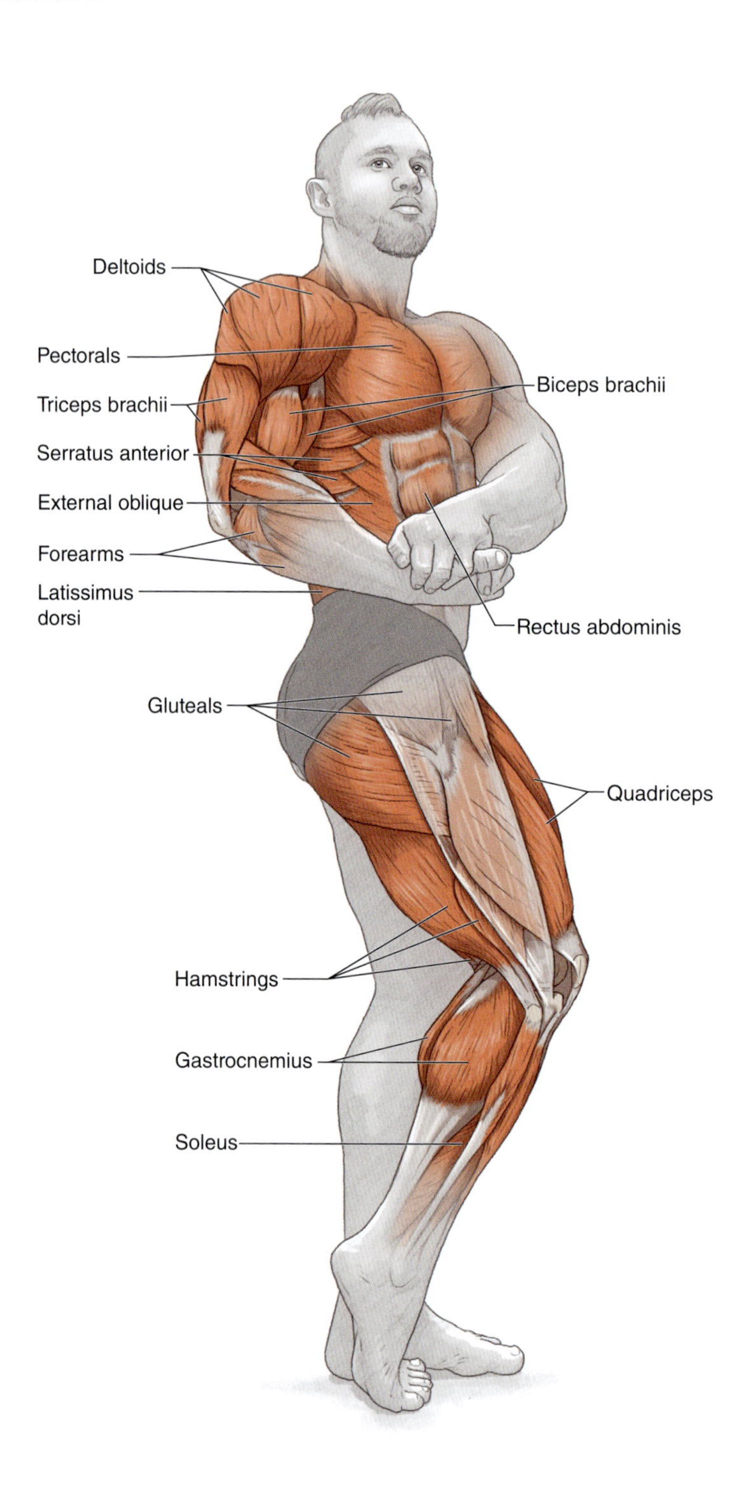
Deltoids
Pectorals
Triceps brachii
Serratus anterior
External oblique
Forearms
Latissimus dorsi
Biceps brachii
Rectus abdominis
Gluteals
Quadriceps
Hamstrings
Gastrocnemius
Soleus

- Front leg about halfway in front of the back leg (some adjustment will be required to get your best look)
- Abs in (either crunched down or vacuum, try both)
- Hips perpendicular to the judges, with the spine rotated as much as possible so that the shoulders are as wide as possible
- Hands clasped at somewhere between hip height or just below chest height (play with various positions to determine your best look)
- Chest up and out as big as possible
- Rear shoulder protected as far as possible to give the biggest illusion of width

Rear Lat Spread

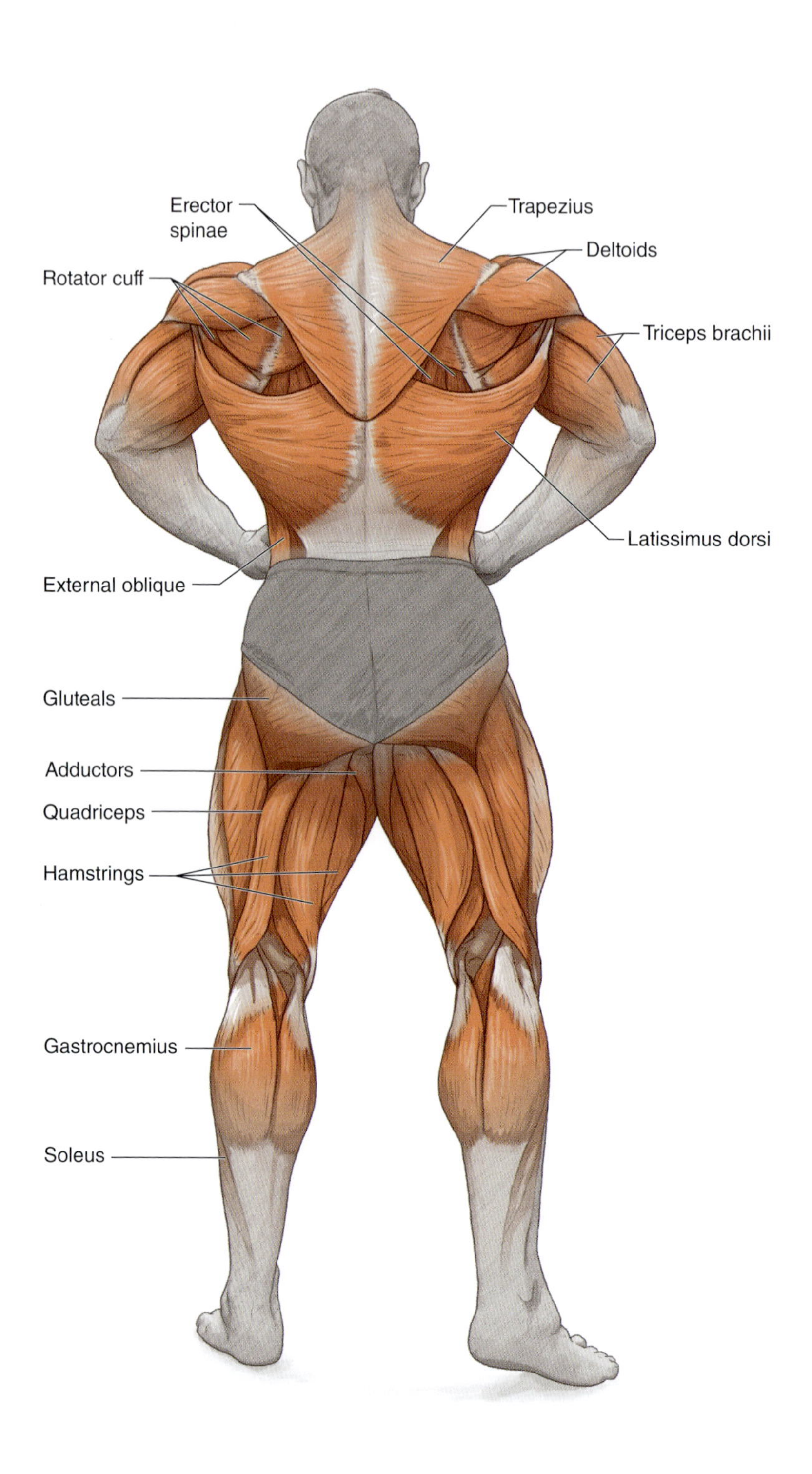

- One foot about four inches (10 centimeters) in front of the midline and the other foot about four inches (10 centimeters) behind it
- Quads flared
- Glutes contracted
- Chest up
- Lean back a lot (hips forward, chest back)
- Shoulders shrugged down
- Pinch the abdomen in from the spine and then, dragging the skin with your thumbs, move the hands forward and *in* to pinch the waist to as small a look as possible
- Lats flared

Rear Double Biceps

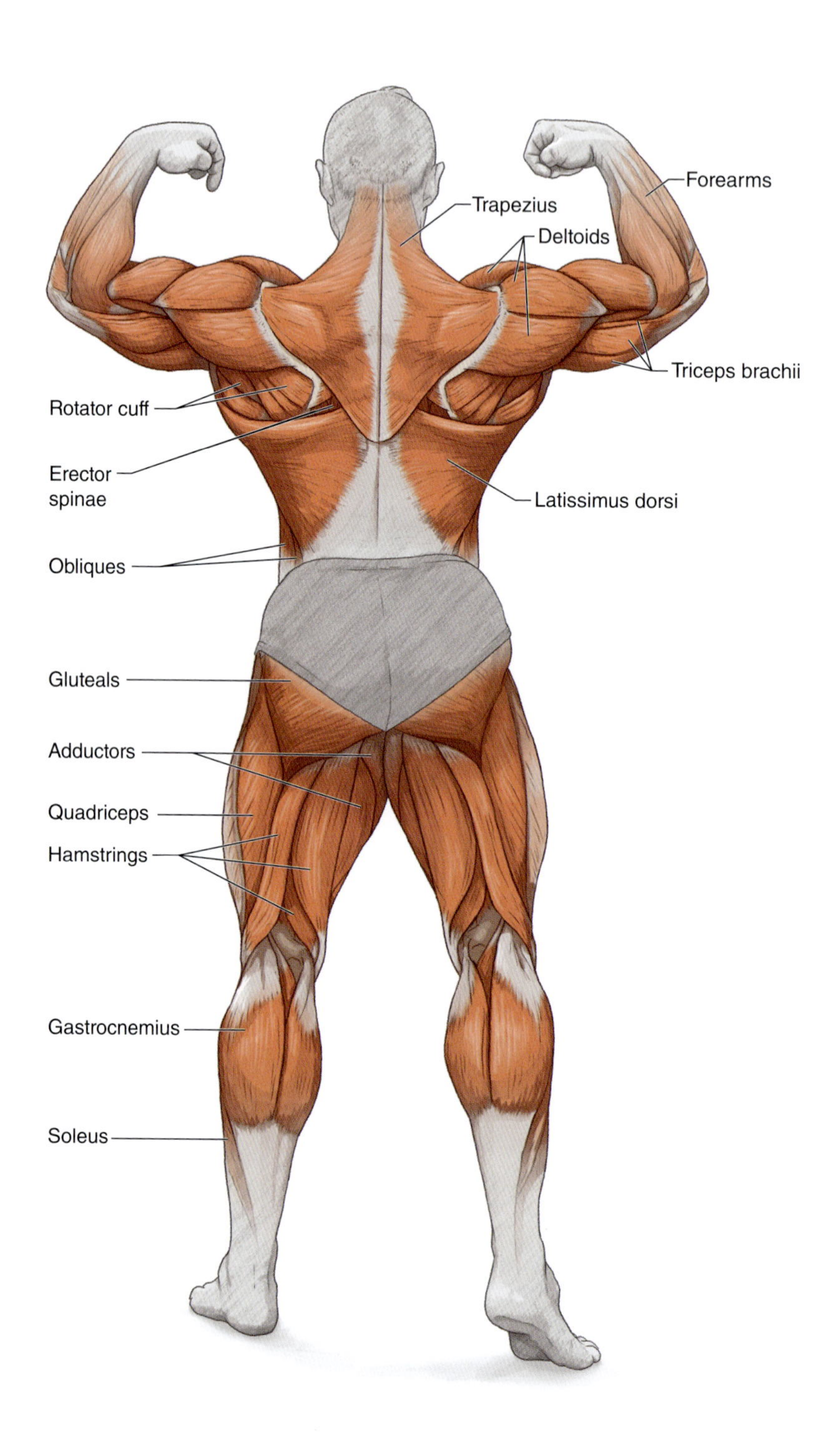

- One foot about four inches (10 centimeters) in front of the midline and the other foot about four inches (10 centimeters) behind it
- Quads flared
- Glutes contracted
- Chest up, with a slight lean backward
- Shoulders shrugged down
- Lats flared
- Elbows *above the line of your shoulders*
- Elbows bent at 70 to 90 degrees (much less or more is a worse look in most cases)
- Elbows rotated forward and wrists rotated into maximum supination (shows off biceps peak best)
- Palms facing toward the head (if they face the rear, the biceps peak won't be as pronounced)

Abdominal and Thigh

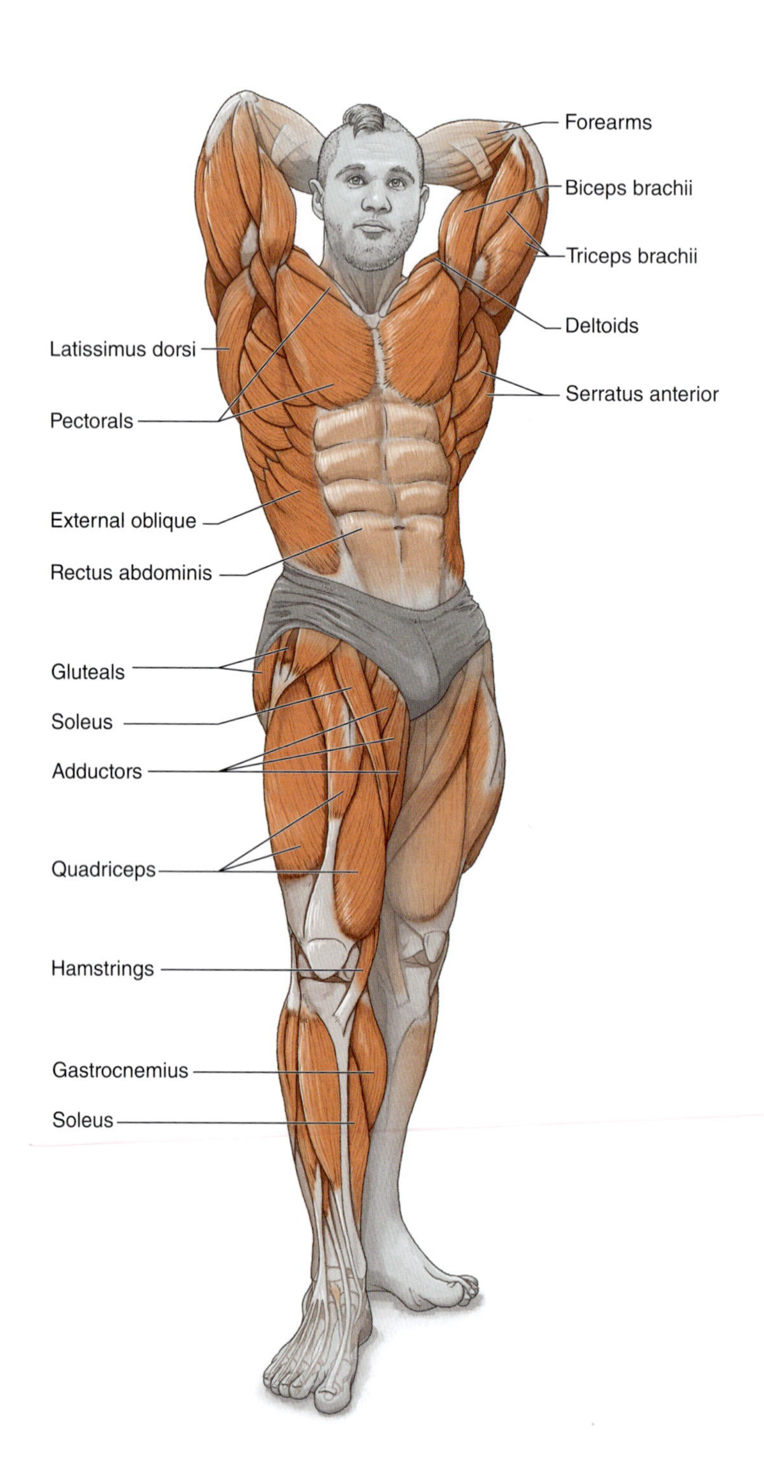

- Quads flared
- Hands behind the head
- Elbows up high and close together
- Abs flexed (a vacuum can work here too)
- Front foot on tippy toes, with weight pulled mostly off the leg and with the knee lifted slightly (toes should still be on the ground, but the knee pull will show off the sartorius and rectus femoris)

Most Muscular

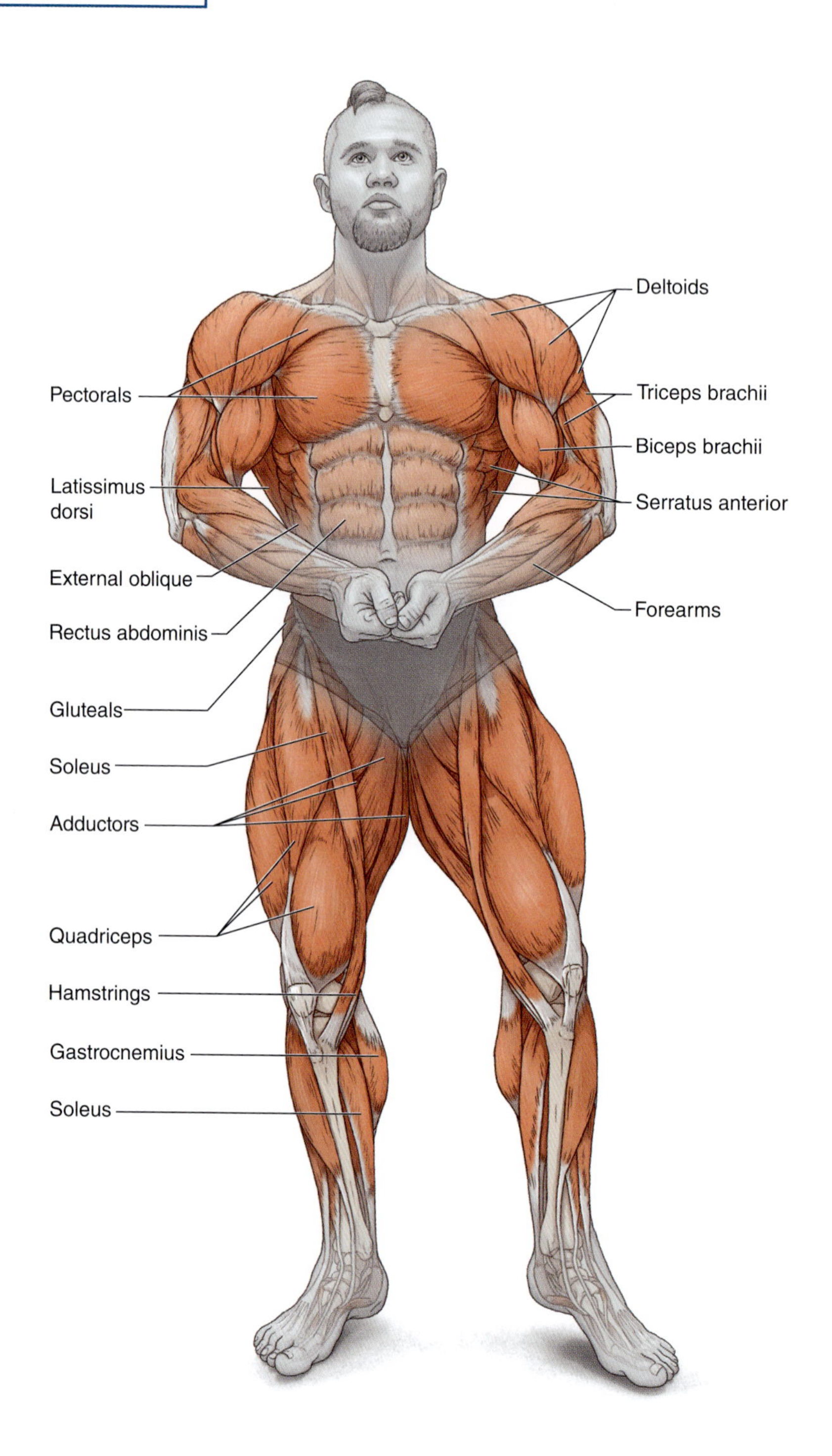

- Quads flared
- Chest up
- Elbows out and forward
- Shoulders shrugged down
- Abs flexed
- Hands clasped, hands on the waist, or the "crab" version (try all three variants to see which ones work best for you)

Conclusion

You can definitely hit any self-created artistic poses in the free posing round; however, in most bodybuilding federations, that round is minimally judged or not judged at all. What is judged is the mandatory posing round. In this round, you have to execute an average of the 12 predetermined poses, with some poses removed and occasionally rarer poses added. All of the competitors hit each pose at the same time so that they can be compared. Each pose follows the aforementioned general and specific posing principles and each one has its own requirements and optimizations *for the individual*. When you learn to pose for bodybuilding, most of the work involves practicing each of those 12 poses to get them as sharp as possible. If you don't practice the standard poses often, you will simply look worse than your muscularity and leanness allow—and if you worked so hard to become muscular and lean, why would you let your posing bring you down at all?

EXERCISE FINDER

Arms 67

Abdominals 107

Legs 123

Posing 181

ABOUT THE AUTHORS

Michael Israetel, PhD, is the cofounder and chief content officer of RP Strength (Renaissance Periodization). With a doctorate in sport physiology from East Tennessee State University, he is also a former professor at Lehman College, Temple University, and the University of Central Missouri, where he taught a variety of exercise science courses related to nutrition as well as strength and hypertrophy.

Israetel worked as a sports nutrition consultant for the U.S. Olympic training site in Johnson City, Tennessee, and has been an invited speaker at numerous scientific and performance conferences worldwide, including nutritional seminars at the U.S. Olympic Training Center in Lake Placid, New York. He has coached numerous athletes in diet and weight training and is also a competitive bodybuilder and professional Brazilian jiujitsu grappler. He is the coauthor of *The Renaissance Diet 2.0*, *Scientific Principles of Hypertrophy Training*, *Scientific Principles of Strength Training*, *Recovering From Training*, *How Much Should I Train?*, and *The Minicut Manual*.

Jared Feather is the head physique and bodybuilding specialist at RP Strength. He holds a master's degree in exercise physiology from the University of Central Missouri. With over 10 years of experience in competitive bodybuilding and coaching, he has earned esteemed titles such as IPE and NFF Professional Natural Bodybuilder. In 2020, Feather placed first in the classic physique division at NPC Nationals, earning him International Fitness and Bodybuilding (IFBB) Pro status. He is also the coauthor of *The Scientific Principles of Hypertrophy Training* and *The Minicut Manual*.

Christle Guevarra, DO, is a former champion powerlifter. She currently maintains a telemedicine private medical practice and serves as an official traveling team physician for U.S. Figure Skating. She is also the production operations manager at RP Strength.

Guevarra holds a bachelor's degree in chemistry and minor in mathematics from San Francisco State University, a master's degree in organic chemistry from University of Wisconsin at Madison, and a doctorate of osteopathic medicine from Western University of Health Sciences. Passionate about education, she was awarded a predoctoral teaching fellowship at Western University of Health Sciences, where she taught first- and second-year osteopathic medical students. She completed her family medicine residency at Crozer Health in Pennsylvania, where she was named Resident of the Year. She then completed her sports medicine fellowship at the University of Nevada at Las Vegas, where she worked with Division I football players, professional athletes, and recreational athletes.

Find more outstanding resources at
US.HumanKinetics.com
Canada.HumanKinetics.com
In the U.S. call 1-800-747-4457
Canada 1-800-465-7301
International 1-217-351-5076
HUMAN KINETICS